Pocket Guide to the Insects of Costa Rica

Pocket Guide
to the
Insects of Costa Rica

Paul E. Hanson
Kenji Nishida
Ángel Solís

Foreword by Tracie Stice

Antlion Media
A Zona Tropical Publication

Comstock Publishing Associates
an imprint of
Cornell University Press
Ithaca and London

First published 2021 by Cornell University Press

Printed in China

Library of Congress Cataloging-in-Publication Data

Names: Hanson, Paul E., author. | Nishida, Kenji, 1972– author. | Solís, Ángel, author. | Stice, Tracie, writer of foreword.
Title: Pocket guide to the insects of Costa Rica / Paul E. Hanson, Kenji Nishida, Ángel Solís, foreword by Tracie Stice.
Description: Ithaca [New York] : Antlion Media, A Zona Tropical Publication; Comstock Publishing Associates, an imprint of Cornell University Press, 2021. | Includes bibliographical references and index.
Identifiers: LCCN 2021003019 | ISBN 9781501760976 (paperback)
Subjects: LCSH: Insects—Costa Rica. | Insects—Costa Rica—Identification.
Classification: LCC QL478.C8 H367 2021 | DDC 595.7097286—dc23
LC record available at https://lccn.loc.gov/2021003019

Zona Tropical Press ISBN 978-1-949469-36-3

Book design: Gabriela Wattson

Contents

Foreword by Tracie Stice vii
Acknowledgments ix

Introduction... 1

Small Orders... 7

True Bugs and Their Kin...........................34

Beetles... 51

Wasps, Bees, and Ants 77

Flies... 101

Butterflies and Moths 112

Other Arthropods... 175

Suggested Reading 189
Photo Credits... 191
Index 193

Foreword

Every insect, spider, or so-called creepy crawly has an incredible story, the bounds of which often defy the imagination. Singing genitalia, fecal shields, and zombie parasites may sound like the inventions of a sci-fi writer but, in truth, there are "aliens" tiptoeing just beyond your front door.

Nothing has given me more pleasure over the past 25 years than stepping into this magnificent realm and sharing the secret lives of insects and their eight-legged kin with travelers from around the world. After a couple of hours of exploration—and with a rekindled sense of wonder—most of my tour guests are not only excited to learn more but are curious to discover the hidden worlds in their own backyards. Insects are everywhere, after all, and there is always something new and exciting to be found.

If each of the earth's currently named insect species could teach a 4-week course covering all the intimate details of its life—how it mates, what it eats, and all the special tricks it has up its chitinous sleeves—it would take more than 80,000 years to complete the curriculum. And consider this: scientists estimate that 80% of all insects have yet to be discovered.

While unraveling these mysteries is a daunting task, there is, thankfully, a dedicated legion of entomologists and enthusiasts alike who have passionately accepted the challenge. In Costa Rica, their numbers include Paul E. Hanson, Kenji Nishida, and Ángel Solís.

If you are captivated by insects and their relatives, then *Pocket Guide to the Insects of Costa Rica* will be an ideal addition to your nature library. Beautiful and easy to use, this guide will introduce you to some of Costa Rica's most charismatic insects, along with a host of other arthropods, through stunning photographs, concise descriptions, and natural history notes.

For travelers, budding entomologists, and guides who wish to expand their knowledge of arthropods, the authors will lead you through the different insect groups, show you where to look and what to look for, and point out the key features to help make identifications. More seasoned naturalists will appreciate the up-to-date diversity statistics specific to Costa Rica and the inclusion of related-species data. Not only is this guide eye candy for the insect enthusiast but it also packs a wealth of information into a field-friendly size.

As you continue seeking out new insect adventures, remember to bring along a sense of curiosity, a keen eye, and *Pocket Guide to the Insects of Costa Rica*. These will stand you in good stead!

Tracie "The Bug Lady" Stice
The Night Tour
Drake Bay, Costa Rica

Acknowledgments

We would like to thank the following people for their invaluable help identifying the insect species shown in photographs and for providing answers to questions: Sabrina Amador, Roberto Cambra, Caroline Chaboo, Oskar Conle, William Eberhard, Bernardo Espinoza, José Antônio Marin Fernandes, Eric Fisher, Bill Haber, Brian Harris, Frank Hennemann, Paul Johnson, Steve Marshall, Juan Mata, Piotr Naskrecki, Dennis Paulson, Jens Prena, Robert Robbins, Bernardo Santos, Monika Springer, Vinton Thompson, and Manuel Zumbado. For identification of some plant species, we are grateful to Mario Blanco, Bill Haber, and Willow Zuchowski. We also thank the following institutions, businesses, and people: Dan Janzen and Winnie Hallwachs (Area de Conservación Guanacaste); Castro-Rodríguez family; Centro Agronómico Tropical de Investigación y Enseñanza (CATIE); Marshall Cobb; El Remanso Rainforest Wildlife Lodge; Marvin Hidalgo (Estación Biológica Monteverde); Finca Café Cristina; Kase Kazuki (Goyi Tours-Casa Mango); Hacienda El Rodeo; Hitoy Cerere Biological Reserve; Jiménez-Calderón family; Erick Berlin (Las Brisas Nature Reserve); Bryna Belisle (Monteverde Butterfly Gardens); Bosque Eterno de los Niños (Monteverde Conservation League); La Selva and Las Alturas Research Stations (Organization for Tropical Studies-Wilson Botanical Gardens); the national parks of Barbilla, Braulio Carillo Sector Quebrada Gonzales, Cahuita, Marino Ballena, Tapantí, and Volcán Tenorio; William and Kristal (Pierella Ecological Gardens); Luis Diego Castillo, Alan Rodriguez, and nature guides (Rainforest Adventures-Tapirus Lodge at Braulio Carrillo); Rancho Naturalista; Rodríguez-Sevilla family; Selva Verde Lodge; Celso Alvarado, Roger Blanco, Fabricio Carbonel, Mario Coto, Javier Guevara, Rafael Gutiérrez, Gustavo Induni, Mariana Jiménez, Alexander León, Isaac López, Alejandro Masís, Miguel Matarrita, Ana María Monge, Henry Ramírez, Lourdes Vargas, and Randall Zamora (Sistema Nacional de Áreas de Conservación-Ministerio de Ambiente y Energía); Tirimbina Biological Reserve; Reserva Biologica Leonelo Oviedo (Universidad de Costa Rica); Laura Chinchilla and Warner Masís (Universidad para la Paz-Zona Protectora El Rodeo); Estación Tropical La Gamba (University of Vienna). We thank Piotr Naskrecki and Aiko Kimura for supplying several stunning photographs (see p. 191). Last but not least, we are grateful to John McCuen of Zona Tropical Press for making this a much better book through his diligent editing.

Introduction

In Costa Rican rainforests—indeed in most terrestrial ecosystems—you are more likely to see an insect than any other animal, so long as you pay attention. While everyone seeks out a view of the large showy butterflies, the majority of insects are smaller—though often every bit as beautiful in their own right—and generally go unnoticed.

Because a small percentage of insects sting, transmit diseases, or destroy our crops, they get a bad rap in the eyes of some. This is a striking irony given that human society and the natural world itself are so dependent on insects. They pollinate most of our food crops and most wild plants. They provide most of the protein for many birds, lizards, frogs, and freshwater fish. Insects facilitate the recycling of decomposing organic matter. And predatory and parasitic insects control the populations of pest insects. But aside from their usefulness, insects include some of the most beautiful, gemlike creatures on the planet.

Insect Anatomy

On insects and their relatives (the non-insect arthropods), the skeleton is located on the outside of the body. Granted, this provides protective armor, but it also means that they must periodically molt (or shed) their exoskeleton in order to grow. Once shed, this exoskeleton is replaced by a larger version. In most insects, this occurs only during the juvenile stages, and adults, therefore, can grow to only a limited extent.

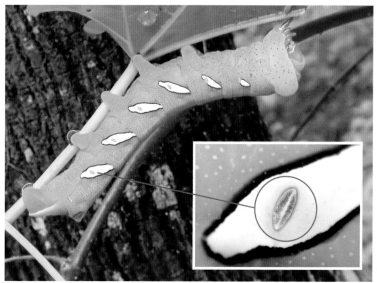

Spiracles in the caterpillar of the Anchemoia Sphinx Moth (*Eumorpha anchemolus*). Red circle indicates location of a spiracle. Insects receive oxygen through these tiny openings.

1

Like most animals, including ourselves, insects have a digestive system, a circulatory system (including a heart), a respiratory system (though they lack lungs), an excretory system, a nervous system (including a brain), and a reproductive system. Insect blood, which is generally not red, flows through the cavities of the body instead of through blood vessels. Rather than relying on blood to distribute oxygen, insects have a network of tiny tubes (tracheae) that carry oxygen directly to all parts of the body. They breathe through tiny holes in the sides of their bodies (the openings to the tracheae, known as spiracles).

We humans rely primarily on vision and sound to perceive the world, but insects rely mainly on their sense of smell; their small size imposes limits on visual and auditory reception. In place of a nose, they smell with their antennae, which have receptors so sensitive that humans have yet to develop artificial detectors to match them. Most communicate with other members of their species via odors known as pheromones. Insects have compound eyes: each eye is made up of numerous independent photoreception units that offer up an extremely pixilated image. While compound eyes generally provide very poor image resolution, they are superb motion detectors. In addition to the set of compound eyes, most adult insects also have one to three tiny, simple eyes (known as ocelli) on the top of their head, which likely help them maintain stability while flying

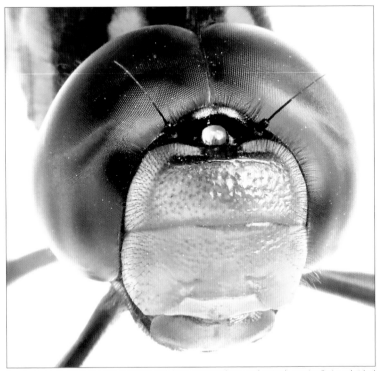

Female dragonfly (*Rhionaeschna cornigera*) with large compound eyes and a simple eye (ocellus) sandwiched between the top of the face and the compound eyes.

Classification

Classifications of organisms are based on hypotheses about their evolutionary history, and as such they are subject to modification when new information becomes available. Thus, recent findings strongly suggest that termites evolved from cockroaches and that lice evolved from barklice. There are 31 orders of insects in the world, 28 of which occur in Costa Rica, and 13 of which appear in this book. Orders are divided into families, families are often divided into subfamilies, then tribes, and ultimately into genera and species.

The insect orders can be arranged into three groups. The most ancient insects appeared before wings evolved and include such wingless insects as springtails and silverfish, though none of these are included in this book. It should be noted that some wingless insects (lice and fleas, for example) evolved from winged ancestors and are therefore grouped with the winged insects. The majority of insects have two pairs of wings; these we can divide into those with an incomplete or gradual metamorphosis, wherein the juvenile stages (nymphs) are quite similar to the adults (except that their wings are not fully grown), and those with complete metamorphosis, in which the juveniles (larvae) look nothing like the adults. Insects with complete metamorphosis must pass through a pupal stage, during which the juvenile body

Developmental stages of the Helenor Blue Morpho (*Morpho helenor*). Left to right, top to bottom: egg, young larva (recently hatched from egg), mature larva, pupa (chrysalis), and adult butterfly.

3

is disassembled and the adult body is formed. The group whose members undergo incomplete metamorphosis includes more insect orders, but the group with complete metamorphosis includes the majority of species because it includes the four largest orders: Coleoptera (beetles), Hymenoptera (wasps, bees, and ants), Diptera (flies and mosquitoes), and Lepidoptera (moths and butterflies).

Insects belong to the class Hexapoda (meaning six legs), which in turn belongs to the phylum Arthropoda (meaning jointed legs). The first chapter (Small Orders), describes a selection of species that do not belong to the "main" orders presented in the subsequent chapters.

Arthopods include three other classes besides the insects and a few examples of these are presented in the final chapter (Other Arthropods). Members of the class Chelicerata are characterized by having four pairs of legs and no antennae; these include spiders, scorpions, and mites. The majority of Myriopoda (among them, centipedes and millipedes) have numerous pairs of legs. Finally, most Crustacea have two pairs of antennae and a variable number of legs, depending on the species; most occur in the ocean, including the animals we are all familiar with: crabs, lobsters, and shrimp. There is now very strong evidence that insects evolved from a group of crustaceans.

About This Book

There is a huge number of insects in Costa Rica—perhaps more than 250,000 species. However many there are, it is estimated more than 80% of them have yet to be named. The species described in this book—a mere drop in the bucket—were selected to give you an idea of some of the colorful and fascinating creatures you will likely encounter on a walk through a rainforest or other habitat in Costa Rica.

For each species we provide one or more photographs, range map (in most cases), and information describing the animal, indicating similar or related species, and conveying natural history snippets. When it was impossible to identify the species in a photograph, only the genus name is given. In a few other cases the entire family (leafhoppers, ichneumon wasps, gall midges) or subfamily (sulfur butterflies and passion-vine butterflies) is discussed and photos of a few representatives are provided.

The organizational structure varies a bit from chapter to chapter. The Small Orders chapter, for example, is divided into orders, while Other Arthropods is divided into subphyla. Most chapters, however, simply present the representative species that belong to the respective group. The Lepidoptera chapter is divided into moths and butterflies, the latter being further divided into families and, in several cases, subfamilies.

Please note that the photos do not show insects at a consistent scale and in fact a given page often shows two different insects at different scales.

Description. Measurements (cm = centimeters, in = inches) of body length do not include antennae or legs. The size of moths and butterflies is given by wingspan (wing tip to wing tip in outstretched wings). In many cases the description and photograph are not sufficient to definitively identify the species. A little known and seemingly bizarre fact about insects is that to identify a species one must often extract and examine the male genitalia or some other microscopic feature.

Related species. This section provides a context for the species under discussion. Obviously, the numbers of species mentioned here are subject to change as more studies are done.

Maps. These are based on data from specimens (collected by the Instituto Nacional de Biodiversidad, now part of the Museo Nacional de Costa Rica); the authors have not been able to verify all the specimen identifications on which the maps are based. As a result, the maps for some species will be subject to modification as new information becomes available.

The authors have tried to avoid using technical terms. When they are used, a brief definition (in parentheses) follows the term. The most essential terms that you will need are shown in the following figure.

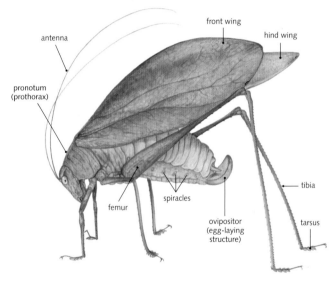

External anatomy of a katydid. The body of an insect is divided into three parts: head (with the eyes and antennae), thorax (with the legs and wings), and abdomen (the rest of the body).

Small
Orders

Dragonflies and Damselflies (order Odonata)

These aerial predators have a unique way of mating. The male uses the tip of his abdomen to hold the female by her neck; she then bends her abdomen forward and upward to receive sperm from the base of the male's abdomen. The immature stages are also predatory, but they live in ponds or streams and breathe with gills. In Costa Rica there are 14 families and around 300 species.

Rubyspot Damselfly (*Hetaerina occisa*)

Description: 4.5 cm (1.8 in) in length. Members of the genus *Hetaerina* are known as rubyspots because the males have a red mark at the base of the wings. The females, which have metallic green marks on the thorax, lack the red spots on the wings. Members of the genus are often difficult to distinguish in the field. **Natural history**: Male rubyspots guard territories along the edges of streams, spending much time perched on the tips of leaves with their body inclined forward. When a female enters his territory, he mates with her (or at least attempts to do so) and continues holding her by the neck even after mating; the male releases the female when she begins laying eggs in submerged plants in the stream. **Related species**: There are 9 species of rubyspot in Costa Rica, the only representatives of the family Calopterygidae in the country.

male

female

Hetaerina occisa

Red-eyed Argia (*Argia cupraurea*)

Description: 4 cm (1.6 in) in length. Males of most *Argia* species have a blue abdomen and can be distinguished from other blue damselflies by their jerky flight; for this reason, they are sometimes called "dancers." They usually perch horizontally with their wings held together and raised slightly above the abdomen. Males of *Argia cupraurea* have red eyes, with the upper part of the thorax metallic red and the abdomen light blue with black rings; females are mostly brownish. **Natural history**: *Argia* species are usually found mostly in open areas, where they capture flying insects. Unlike other genera in the family Coenagrionidae, the immature stages are found in running water instead of ponds. Males often perch on sun-bathed rocks in rivers. The male continues to grasp the female as she lays eggs in algae growing on submerged rocks. **Related species**: *Argia* is the largest genus (nearly 30 species in Costa Rica) in the largest family (Coenagrionidae) of damselflies. *Argia calverti* and *A. oenea* also have red-eyed males and can be difficult to distinguish from *A. cupraurea* in the field. *A. anceps* and *A. elongata*, which lack red eyes in the males, are abundant in upland regions.

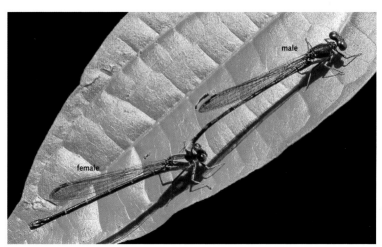

Argia cupraurea. Male holding female with claspers.

Argia anceps breeds more frequently in ponds than in streams.

Giant Helicopter Damselfly (*Megaloprepus caerulatus*)

Description: 8–11 cm (3.1–4.3 in) in length. This species has the largest wingspan of any damselfly or dragonfly in the world, up to 19 cm (7.5 in). In addition to its size, note the large black band near the tip of each wing; at least on the Caribbean slope, males are easily distinguished from females by having a large white area preceding the black band. Slow flight combined with the whirling movement of the colored wing tips gives the illusion of miniature helicopter blades. Helicopter damselflies perch by hanging from vegetation. **Natural history**: The adults prey on insects that are sitting on leaves and sometimes snatch spiders or trapped insects from spider webs. Males maintain territories in light gaps of forests, where they drive away other males and wait for a female to arrive. After mating, the female lays her eggs in water contained in a nearby tree hole. The immature stages (nymphs) feed on animals inhabiting the water, including other aquatic insects and tadpoles. The nymphs require several months to reach adulthood. **Related species**: Helicopter damselflies (3 genera in the family Coenagrionidae) include 5 species in Costa Rica. The other species are slightly smaller and differ in color.

Megaloprepus caerulatus. Male in flight (above). Resting male (left).

Turquoise-tipped Darner
(*Rhionaeschna psilus*)

Description: 6 cm (2.4 in) in length. The sexes are quite similar, though males often have blue eyes. This dragonfly has two broad, pale green stripes on the sides of the thorax and a series of small green markings all along the abdomen. **Natural history**: A very common species that sometimes even enters buildings. The males fly back and forth along the edges of ponds searching for females. Females lay their eggs on damp ground near the water's edge or on floating rotten wood. **Related species**: Two other similarly colored species of this genus occur in Costa Rica, *R. cornigera* being even more common than *R. psilus*. They are members of the family Aeshnidae.

Female *Rhionaeschna psilus*

Lowland Knobtail (*Epigomphus tumefactus*)

Description: 5 cm (2 in) in length. Unlike other dragonflies, members of this family (Gomphidae) have widely separated eyes (when viewed from above). They also differ by flying slightly slower and landing immediately on a perch—without hovering a few seconds before setting down. The sexes are quite similar, although the tip of the male's abdomen is somewhat enlarged, hence the name *knobtail*. Knobtails have a dark brown thorax with light yellow stripes and a black abdomen with well-spaced, small whitish markings. **Natural history**: Knobtails spend prolonged periods of time perched in a horizontal position on leaves, branches, or rocks. From this perch they dart out to capture flying insects. Males perch at the edge of a stream but do not make frequent patrolling flights as many other dragonflies do. They mate for an hour or longer while perched on vegetation. The male does not accompany the female when she goes to lay eggs in the water. **Related species**: Costa Rica has 13 species of *Epigomphus*; it is the largest genus in the family Gomphidae.

Epigomphus tumefactus. Female in light gap of forest understory.

Silver-sided Skimmer (*Libellula herculea*)

Description: 5 cm (2 in) in length. The thorax is dark brown, with a white (male) or yellow (female) stripe running down the middle of the top surface; powdery gray on the sides. The broad abdomen is a deep crimson-red in males and dark orange in females. They perch with the body tilted slightly downward. **Natural history**: Found near ponds and sections of still water in streams. Males guard territories near shaded ponds or occasionally sunlit pools of water, though they spend most of their time perched on branches. Mating is brief, but the male remains flying near the female as she lays eggs. She does this by hovering over the water and using the tip of her abdomen to scoop up a drop and flicking it, together with some eggs, onto the shore. **Related species**: There are 4 species of *Libellula* in Costa Rica; they belong to the family Libellulidae.

Libellula herculea. Male perching along a road.

Roseate Skimmer (*Orthemis ferruginea*)

Description: 5 cm (2 in) in length. Males are almost entirely pinkish-purple; females have a brown thorax, with a light-colored stripe running down the middle of the back and an orangish-brown abdomen. **Natural history**: Because the immature stages inhabit a wide diversity of freshwater habitats, including even polluted waters in urban areas, this is one of the most common dragonflies in the country. Males sit for brief periods on branches near water but frequently leave their perch to chase away other males. If a female enters his territory, the male approaches her with his abdomen lowered. After a brief period of mating, the female hovers over the water while using the tip of her abdomen to scoop up a drop and flick it, together with some eggs, onto the shore. **Related species**: There are 7 species of *Orthemis* in Costa Rica; they belong to the family Libellulidae.

Male *Orthemis ferruginea*

Pond Amberwing (*Perithemis tenera = P. mooma*)

Description: 2.5 cm (1 in) in length. The smallest dragonflies, amberwings (genus *Perithemis*) are named for the yellowish-orange color on the wings of the male. The stout body is also yellowish-orange. **Natural history**: Males perch on objects just above the water and chase other males out of their territory. When a female enters his territory, the male courts her by flying in front of her, vibrating his wings faster than normal, and raising his abdomen (such courtship displays are rare in dragonflies). The male then leads the female to a place to lay eggs; if she approves of the spot, they briefly mate and the female then lays her eggs as the male remains nearby. Eggs are laid on submerged wood sticking out of the water or on floating vegetation. **Related species**: There are 3 species of amberwing in Costa Rica; they belong to the family Libellulidae.

Male *Perithemis tenera*, dorsal view.

Male *Perithemis tenera*, lateral view.

14

Earwigs are easily recognized by large pincers at their rear end and very short front wings, which conceal the well-developed, membranous hind wings used for flying (a few species lack wings). In Costa Rica, there are 6 families and around 100 species of earwig.

Giant Earwig (*Carcinophora americana*)

Description: 3–4 cm (1.2–1.6 in) in length. Large and sporting bright orange front wings, this is one of the more striking earwigs in Costa Rica. **Natural history**: Earwigs are unduly feared, probably because of their formidable pincers, but in fact they are harmless. The pincers are used to fold and unfold the hind wings, to capture prey, and in self-defense. Earwigs usually hide in leaf litter, rotting tree trunks, and similar sites during the day; they are thought to be omnivorous, with predatory tendencies. The female builds a nest, usually in the ground, and cares for her eggs, licking them to protect them from fungal infection. **Related species**: By current count, there are 5 species of *Carcinophora* in Costa Rica; they belong to the family Anisolabididae.

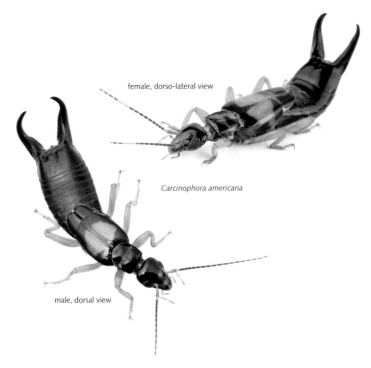

female, dorso-lateral view

Carcinophora americana

male, dorsal view

Relative to body size, male pincers are typically larger than those of females.

Katydids, Crickets, and Grasshoppers (order Orthoptera)

Most members of this group have enlarged hind legs that are used for jumping. Species are divided into two groups or suborders. Katydids and crickets generally have very long antennae and are mostly nocturnal, whereas grasshoppers have shorter antennae and are diurnal. Males of many katydids and crickets sing, which they do by rubbing their front wings together. Relatively few grasshoppers sing, but those that do create sounds by rubbing the front wing against the hind leg or against the hind wing. In Costa Rica there are 16 families and well over 600 species.

Sylvan Leaf Katydid (*Mimetica mortuifolia*)

Description: 5 cm (2 in) in length. Species in this genus have front wings that resemble dead or partially dead leaves, complete with wing venation that simulates leaf veins. The hind wings are reduced. **Natural history**: As the name *Mimetica* suggests, these katydids are incredibly good leaf mimics. What's more, even within a given species, nearly all individuals differ—some resemble a green leaf, others a damaged leaf, still others a dead leaf. This is to prevent monkeys, one of their main predators, from easily learning what they look like. At night they feed on foliage and males make short buzzing calls to attract females. **Related species**: There are at least 5 species of *Mimetica* in Costa Rica. They belong to the subfamily Pterochrozinae (family Tettigoniidae).

Female *Mimetica mortuifolia*

Another species of *Mimetica* (a female)

Yellow-faced Spear Bearer (*Copiphora cultricornis*)

Description: 6–7.5 cm (2.4–3 in) in length. Like many other members of the subfamily Conocephalinae (meaning "conehead"), these katydids have a horn on the top of their head. This species is green with a yellow face that shows 4 or 6 small black dots. Females have an extremely long, straight ovipositor (egg-laying organ) projecting from the rear end, hence the name *spear bearer*. **Natural history**: Like many other katydids, this species is nocturnal. Not much is known about its diet, but it appears to feed on fruits, seeds, and slow moving insects. The male's song consists of a widely spaced series of chirps. Females probably come down to the ground to lay eggs. **Related species**: There are at least 6 species of *Copiphora* in Costa Rica. They belong to the subfamily Conocephalinae (family Tettigoniidae).

Female *Copiphora cultricornis*, lateral view.

Female *Copiphora cultricornis*, frontal view.

17

Champion's Katydid (*Championica montana*)

Description: 5.5–6.5 cm (2.2–2.6 in) in length. This species is an incredibly good mimic of moss, not only in its coloration but also due to the spines projecting from its body and legs. **Natural history**: If confronted with a potential predator, this katydid spreads its wings and kicks with its spiny hind legs. Males of another species in the genus were found singing an hour after midnight, producing a succession of readily audible notes. **Related species**: There is at least one other species of this genus in Costa Rica, *C. cristulata*. Both belong to the subfamily Pseudophyllinae (family Tettigoniidae).

Female *Championica montana*, lateral view.

Frontal view of *Championica montana*.

18

Broad Wing Katydid (*Stilpnochlora azteca*)

map for genus

Description: 6–7 cm (2.4–2.8 in) in length. This species is entirely green except for a dark band on the upper thorax (pronotum). The wings resemble a leaf; the hind legs are noticeably longer than the other legs and are used for jumping. As in other members of the subfamily (Phaneropterinae), the tips of the hind (inner) wings protrude slightly beyond the front wings, and the female's egg-laying organ (ovipositor) is relatively short, flattened from side to side, and curved upward. **Natural history**: This species is nocturnal and feeds on the leaves of trees. The female attaches her flat, oval-shaped eggs to a twig, where the eggs are arranged like shingles on a roof; they require more than a month to hatch. **Related species**: There are at least 4 species of *Stilpnochlora* in Costa Rica. They belong to the subfamily Phaneropterinae (family Tettigoniidae).

Stilpnochlora azteca feeding on leaf of *Montanoa guatemalensis* (Asteraceae).

Monkey Grasshopper (*Homeomastax surda*)

Description: 1 cm (0.4 in) in length. Relatively small, Monkey Grasshoppers (family Eumastacidae) are colorful and have very short antennae, shorter than the front femur, with 14 or fewer segments. They can be recognized by the fact that their hind legs stick out, perpendicular to the body. The family Episactidae is quite similar but the adults are wingless. **Natural history**: These grasshoppers occur in wet forests, where they feed on ferns and a variety of other plants. They are thought to insert their eggs into plant tissue. **Related species**: There are 10 species of Eumastacidae in Costa Rica, all in the genus *Homeomastax*. The only one that occurs in the same areas as *H. surda* is *H. bustum*, which appears to be less common; on *H. bustum*, the lower face is pale yellow (bluish on *H. surda*). The family Episactidae is represented in the country by just a single species, *Episactus tristani*, which occurs at mid-elevations, from Guanacaste to the Central Valley.

Male *Homeomastax surda*, dorsal view. Note how the hind femora are held perpendicular to the body.

Proctolabine Grasshopper (*Ampelophilus truncatus*)

Description: 1.5 cm (0.6 in) in length. Adults have reduced wings. Body mostly green but with considerable blue on the head and with a black band behind the eyes; hind knees dark above and blue below; hind tibiae blue. In males, the tip of the abdomen is reddish (reddish-brown in female). **Natural history**: Species in this genus feed on plants in the nightshade and aster families (Solanaceae and Asteraceae respectively). *A. truncatus* is often found on *Acnistus arborescens*, a small very common tree in the nightshade family. **Related species**: The genus *Ampelophilus*, comprising 4 species in Costa Rica, belongs to the tropical American subfamily Proctolabinae (family Acrididae).

Female *Ampelophilus truncatus*

Bird Grasshopper (*Schistocerca centralis*)

Description: 8 cm (3.1 in) in length. This large grasshopper is light brown, with a pale stripe running down the center of the back. **Natural history**: Some members of this genus are locusts, which means that they form swarms when conditions are crowded. *Schistocerca centralis*, however, is a normal (solitary) grasshopper that prefers somewhat drier habitats. There is little information about its feeding preferences, but nymphs and adults perhaps utilize different plants. **Related species**: There are 4 species of *Schistocerca* in Costa Rica, one of which (*S. piceifrons*) occasionally forms swarms farther north in Central America. In Costa Rica it occurs only in the northwestern part of the country, where it is solitary. In total there are nearly 40 species, all found in the Americas except for the Desert Locust (*S. gregaria*) of biblical fame. *Schistocerca* is the only genus in the subfamily Cyrtacanthacridinae (family Acrididae) that occurs on the mainland of the Americas.

Schistocerca centralis. A few species in this genus are migratory locusts that occasionally devastate crops.

Reticulated Lubber Grasshopper (*Taeniopoda reticulata*)

Description: 5–7 cm (2–2.8 in) in length. This species is quite easily recognized by its relatively large size, black and dark purplish coloration (though colors vary), including purplish reticulations in the front wings, and yellow antennae with black tips. The top of the thorax is crenulated and prolonged backward. The gregarious nymphs are black with red lines on the head and another red line along the top and back edge of the thorax. **Natural history**: This species feeds on a wide variety of plants. It tends to seek refuge in a shady spot during the middle of the day and feeds mostly from evening until the following morning. If disturbed, it raises its front wings to expose the red hind wings; if further disturbed, it emits a frothy substance and hissing sound from its thoracic spiracles (respiratory openings). **Related species**: The only other Costa Rican species in the genus, *T. varipennis*, is more common on the Pacific side and is dark green. There are about 40 species of lubber grasshoppers (family Romaleidae) in Costa Rica; one of these, the Giant Red-winged Grasshopper (*Tropidacris cristata*), is the largest grasshopper (14 cm / 5.5 in) in the country. Lubber grasshoppers occur only in the Americas, unlike the cosmopolitan Acrididae, the family to which most grasshoppers belong.

Taeniopoda reticulata mating, with male on top of female. Frontal view (inset).

23

Stick Insects / Walking Sticks (order Phasmatodea)

The name for this order contains the root word for phantom, appropriate enough for insects that are so adept at blending in with their background. They feed mostly at night, on the leaves of trees and shrubs. There are 5 families and at least 80 species of stick insect in Costa Rica.

Moss Stick Insect (*Trychopeplus laciniatus*)

Description: About 7–9 cm (2.8–3.5 in) in length, although the body is usually curved. Color varies from green to brownish. The females are wingless; adult males have wings and are slightly smaller. Both the body and legs resemble a moss-covered twig; the camouflage is enhanced by this insect's habit of swaying its body when the wind blows. **Natural history**: This species feeds mostly on orchids but has also been observed feeding on *Schefflera* and *Smilax*. It feeds primarily during the day; at night it hangs motionless from vegetation, probably to avoid nocturnal predators. The female flicks her eggs away from her body; the eggs drop to the ground or vegetation below. Interestingly, her brown eggs are covered with an elaborate arrangement of small spines. **Related species**: This is the only species in the genus currently known from Costa Rica. It belongs to the family Diapheromeridae, which has at least 20 species in the country.

Trychopeplus laciniatus is a master of camouflage; note this female hidden in area indicated by the circle.

24

Mating *Tychopeplus laciniatus*; male on left, female on right feeding on *Smilax* leaf (Smilacaceae).

Trychopeplus laciniatus eggs (about 0.5 cm / 0.2 in long).

Phanocles Stick Insect (*Phanocles costaricensis*)

Description: Females 19 cm (7.5 in) in length; males 12 cm (4.7 in). The body is extremely long and slender, uniformly grayish brown in color, and devoid of wings. **Natural history**: This species feeds mostly at night, on the leaves of various plants. The female flicks her eggs away, and they drop to the ground or vegetation below. **Related species**: There are at least 4 species of this genus in Costa Rica, and they are all quite similar. They belong to the family Diapheromeridae.

Female feeding on *Rondeletia* (Rubiaceae).

Egg of *Phanocles* sp. (about 0.5 cm / 0.2 in long).

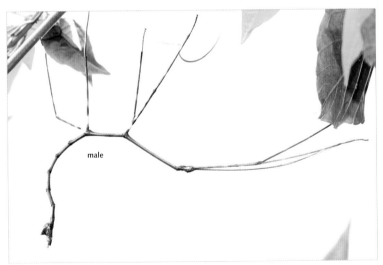

Phanocles costaricensis, one of the largest stick insects in Costa Rica.

Mantises (order Mantodea)

Mantises are close relatives of cockroaches, but unlike their detritis-eating cousins, all mantises are predators. Although it is often said that the female mantis eats the male after copulating with him, the male usually does his best to avoid this outcome—and he is usually successful. There are 6 families and at least 80 species in Costa Rica.

Bark Mantis (*Liturgusa maya*)

Description: 3 cm (1.2 in) in length. Body quite flat, with green and black camouflage mottling. The legs have green and black bands; the middle and hind legs are relatively long and extend outward from the side of the body. The front part of the thorax (pronotum) is about 3 times as long as it is broad. **Natural history**: Bark Mantises (family Liturgusidae) inhabit the bark of trees, especially those with relatively smooth, lichen-covered bark. They are usually found with the head facing down. During the day they often scurry around to the other side of the tree when approached and are more easily seen at night with a flashlight. The egg case (ootheca) is spherical with a long tubular neck (through which the hatching nymphs exit). **Related species**: There are at least 3 species of *Liturgusa* in Costa Rica; these and one species of *Corticomantis* are the only members of the family Liturgusidae in the country.

Liturgusa maya

Egg case (ootheca) of *Liturgusa maya*, with young emerging from top.

27

Shield Mantis (*Choeradodis rhombicollis*)

map for genus

Description: 7 cm (2.8 in) long. Praying mantises belonging to the genus *Choeradodis* are known as shield mantises because of their greatly expanded and flattened thorax (pronotum), which is larger and more square-shaped in females and more triangular in males. They are leaflike in appearance and are therefore also known as leaf mantises. **Natural history**: The camouflaged body not only protects this species from predators but also allows it to go unnoticed by passing prey. Adults, especially females, can live for more than a year, time enough for their body to become colonized by various species of lichens and liverworts. The female lays her eggs in a large, oval-shaped packet (ootheca) on vegetation. **Related species**: There is at least one other species of *Choeradodis* in Costa Rica, *C. rhomboidea*; females of the latter species have a pronotum that is less square-shaped. This genus is the only member of the subfamily Choeradodinae (family Mantidae) in the country.

Choeradodis rhombicollis. Camouflage makes it easier to ambush prey and hide from birds and other predators.

28

Horned Mantis (*Vates pectinicornis*)

Description: 6–8 cm (2.4–3.1 in) in length. As a group, horned mantises (subfamily Vatinae) usually have a prominent horn on top of the head (they are also known as unicorn mantises). The middle and hind legs have protruding lobes that probably contribute to camouflage. The genus *Vates* can be distinguished by its thorax (pronotum), which has a smooth surface and a small lateral expansion on each side near the front end. **Natural history**: Like many mantises, this species utilizes a strategy of waiting for its prey, then ambushing it. When threatened it becomes a stick mimic, freezing in a semi-upright position. **Related species**: There are only two genera of Vatinae (family Mantidae) in Costa Rica. *Vates*, with 3 species, and *Pseudovates*, with at least 2 (probably more) species in the country.

Vates pectinicornis and (inset) frontal view of head of *Pseudovates chlorophaea*.

Cockroaches and Termites (order Blattodea)

The half a dozen species of cockroach that have cohabitated with humans around the world have given the entire group a bad reputation. However, the vast majority of species, at least 200 in Costa Rica (in 5 families), live only in the forest. Up until recently termites were classified in a separate order (Isoptera), but there is now strong evidence that they evolved from cockroaches. Like ants, all termites are eusocial, i.e. they live in colonies consisting of a queen and large numbers of workers; but unlike ants, termite colonies also have a king. There are 3 families and about 80 species of termite in the country

Green Banana Cockroach (*Panchlora nivea*)

Description: Males 1.2 cm (0.5 in) in length; females 2.5 cm (1 in). Readily recognizable by its pale green color. The nymphs are dark brown and seen less often due to their more secretive habits. **Natural history**: This species is attracted to lights. The nymphs feed in rotting logs, including those of fallen coconut trees, and in decomposing banana stalks. The diet of adults is poorly known, but indirect evidence suggests that they feed on nectar in the canopy. Unlike some cockroaches, which deposit their egg case (ootheca) in the environment, the Green Banana Cockroach carries the egg case inside her body until the eggs hatch. **Related species**: There are at least 6 species of *Panchlora* in the country. They belong to the family Blaberidae.

Panchlora nivea, not your stereotypical cockroach.

Nasute Termite (*Nasutitermes corniger*)

Description: Workers and soldiers roughly 0.6 cm (0.2 in) in length. The nests of *Nasutitermes* are commonly seen in trees or on fence posts; they are round to elliptical in shape, often larger than a basketball, and dark brown with a bumpy surface. Trails covered with dark debris lead away from the nest to where the termites are feeding. While the workers and winged adults resemble most other termites, the soldiers can be identified by their large, shiny heads that have a cone-shaped snout (in Latin, *nasutus* means "large-nosed") and by the fact that they lack mandibles entirely. **Natural history**: Like other termites, *Nasutitermes* live in colonies consisting of workers, soldiers, a queen, and a king. Only queen and king reproduce. The soldiers protect the colony from enemies such as ants and anteaters by using their snout to spray a sticky repellent that smells like turpentine. The workers bring food to the colony, primarily dead wood and other sources of cellulose, which they digest with help from microbes living in their guts. A mature colony can have as many as 900,000 workers and once a year the colony produces from 5,000 to 25,000 winged males (new kings) and females (new queens), which shed their wings, mate, and begin new colonies. New colonies often have multiple queens and kings all living in the same royal chamber, but eventually most colonies have just a single functioning queen and king. **Related species**: There are at least 5 species of *Nasutitermes* in the country. They belong to the largest subfamily (Nasutitermitinae) of the largest family (Termitidae).

Nasutitermes corniger, workers and soldiers

Nest on a large tree trunk.

Nest on a branch high above ground.

Dobsonflies (order Megaloptera)

Members of this small group have aquatic larvae. There are just two families in Costa Rica: Corydalidae, with a dozen species, and Sialidae, with just a single, rarely seen, species.

Dobsonfly (*Corydalus* species)

map for genus

Description: 8–12 cm (3.1–4.7 in) in length. These very large, brownish insects hold their wings flat over the body. Male *Coydalus* are easily recognized by their extremely long mandibles. While these males look ferocious, they are actually quite harmless and one hardly feels the "bite." Although females have a mandible that is less intimidating, they can bite very hard if handled. **Natural history**: Adult dobsonflies are nocturnal and often come to lights. They eat very little; males probably subsist on nothing but water, and females are thought to consume only nectar or fruit juice, depending on availability. Males use their mandible in fights with other males and in courting females. The latter lay their eggs in a large cluster (sometimes more than 1000 in a single cluster) that is encased in a white protective material and attached to plants or rocks near streams. Each female deposits two or three egg masses during her short lifetime. The larvae (known as hellgrammites) are voracious predators of other aquatic insects; when mature, they leave the water to pupate either in soil under rocks or in rotting logs. **Related species**: There are 7 species of *Corydalus* in Costa Rica; they belong to the family Corydalidae.

Corydalus species

Caddisflies (order Trichoptera)

Caddisflies are close relatives of moths, though their wings are covered with hairs instead of scales; in addition, caddisflies often have longer antennae. The larvae live in streams or ponds; several species build a portable case around their body. There are 15 families and nearly 500 species of caddisfly in Costa Rica.

Net-spinning Caddisfly (*Leptonema* species)

map for genus

Description: About 2–3 cm (0.8–1.2 in) in length, not counting the very long, threadlike antennae. Most species in this genus are light green. **Natural history**: Like many caddisflies, *Leptonema* species are nocturnal and attracted to lights. The larvae live in streams, where they build silken nets with which to capture organic particles carried by the current. **Related species**: Net-spinning caddisflies belong to the family Hydropsychidae; the largest genus in this family is *Leptonema*, with about 30 species in Costa Rica. Most species look quite similar and can only be distinguished by characteristics of the males' genitalia.

Most species of *Leptonema* are light green, but some, including this species, are light brown.

True Bugs and Their Kin

(order Hemiptera)

True Bugs and Their Kin (order Hemiptera)

True bugs have needle-like or threadlike mouthparts adapted for piercing and for sucking liquid food, from other insects or from plants. There are three groups (suborders): the Heteroptera (true bugs), Auchenorrhyncha (cicadas, planthoppers, treehoppers, leafhoppers), and Sternorrhyncha (white flies, aphids, scale insects, mealybugs, etc.).

Wheel Bug (*Arilus gallus*)

Description: Nearly 4 cm (1.6 in) in length. Species in the genus *Arilus* are known as wheel bugs and are easily recognized by the semicircular crest (which looks like a cogwheel) at the front end of their thorax. **Natural history**: Like most other bugs known collectively as assassin bugs, this species is a predator of other insects; it is most commonly found on trees and shrubs. Rather slow moving, it preys primarily on soft-bodied insects such as caterpillars. After capturing an insect with its front legs, the wheel bug stabs it with its beak and injects it with enzymes that paralyze the prey and dissolve its innards. The wheel bug then slowly sucks out the liquefied contents of its victim. After mating, the female glues small clusters of eggs to the surface of a plant. **Related species:** This is the only species of the genus reported from Costa Rica. In total there are about 300 species of assassin bugs (family Reduviidae) in the country.

Arilus gallus. Like other assassin bugs, this species has a stout beak (below head) that is used for stabbing prey. Like other true bugs it emits a defensive odor; this species gives off the smell of sweet sulfur.

Bee Assassin (*Apiomerus vexillarius*)

Description: 2.5 cm (1 in) long. Bee assassins (genus *Apiomerus*) have a rather wide body, often black with yellow or red at the base of the wings; the front and middle legs appear to have stubby endings. Females of this species have a pair of bright red, circular appendages (foliaceous appendages) projecting from their rear end (their function is unknown). **Natural history**: Although they prey on a variety of insects, bee assassins often wait on flowers or the nest entrances of bees (especially stingless bees) to capture bees. They appear to emit a chemical substance, perhaps smelling like a flower, which attracts the bees. Bee assassins also have the unusual habit of coating their front legs with sticky plant resins that facilitate grabbing their prey and holding it as they stab it with their beak. Females of some species also store resin on the abdomen, which they use in coating their eggs in order to protect them from predators such as ants. **Related species**: There are about 10 species of *Apiomerus* in Costa Rica. They belong to the family Reduviidae.

Female *Apiomerus vexillarius*

Apiomerus hirtipes, found in the Cordillera de Guanacaste and the Caribbean slope below 1000 m (3280 ft).

Kissing Bug (*Triatoma dimidiata*)

Description: 3 cm (1.1 in) long. *Triatoma dimidiata* can be distinguished from nearly all other kissing bugs by its black thorax and an abdomen whose sides are orangish colored with a black circular spot on each segment. A few leaf-footed bugs have similar coloration, but their head is not as long as that of kissing bugs. **Natural history**: Kissing bugs suck blood from vertebrates. Many feed only on the blood of wild mammals or birds, but others such as *T. dimidiata* are generalists. This species hides during the day, often in wood piles, and becomes active at night. In Central America it is the main vector of Chagas disease, although only about a third of the bugs carry the protozoan causing this disease. The protozoan (*Trypanosoma cruzi*) is transmitted not by the bite but rather in the feces, which people can inadvertently rub into their mouth or eyes while sleeping. Although this is a very serious disease in the Americas, kissing bugs are generally absent in modern housing and hotels. **Related species**: Costa Rica has 12 species of kissing bug (subfamily Triatominae, family Reduviidae), 4 of which belong to the genus *Triatoma*.

nymph

Triatoma dimidiata adult on floor and late stage nymph on human arm (inset).

Bullhorn Stink Bug (*Edessa tauriformis*)

Description: 2 cm (0.8 in) long. The scientific name of this species is derived from *tauros*, which is Greek for bull, appropriate since it has a very long black horn on each side of the front of the body (pronotum). **Natural history**: Like most stink bugs, this species feeds on plants and emits a strong pungent odor when it is disturbed. Males and females probably communicate with one another by vibrating the plant surface. Females of another species in this genus have been observed laying eggs at night, in a mass consisting of 2 rows of about 7 eggs per row, usually on the under surface of the leaf. **Related species**: There are 56 species of *Edessa* in Costa Rica, by far the largest genus in the subfamily Edessinae. In total there are probably 300 species of stink bug (family Pentatomidae) in the country.

Edessa tauriformis in the understory of a rainforest.

Shield Bug (*Pachycoris torridus*)

Description: 1.5 cm (0.6 in) long. Members of the family Scutellaridae are known as shield bugs because of their large, convex scutellum (most visible part of the thorax), which completely covers the wings and abdomen, giving them a beetle-like appearance. *Pachycoris torridus* comes in several different colors but most individuals have a black background with a dozen red or yellow spots; the number and shape of the spots is variable, however, as is the background color. **Natural history**: Species in this genus generally feed on plants in the spurge family (Euphorbiaceae). The female stands guard over the egg mass as well as the very young nymphs. Young nymphs remain grouped together, but as they grow older they begin dispersing. The bright coloration of the nymphs and adults probably serves as a warning to potential predators that these insects contain chemical defenses, which they obtain from the host plant. **Related species**: There are about 30 species of shield bug in Costa Rica, at least two of which belong to the genus *Pachycoris*.

male in flight, ventral view

Pachycoris torridus male. In shield bugs the wings are completely concealed.

Leaf-footed Bug (*Anisoscelis alipes*, previously *A. flavolineatus*)

Description: 2 cm (0.8 in) long. Like many other members of the family Coreidae, this species has a striking leaflike expansion on the hind legs. The head and front part of the thorax are light orange while the rest of the top surface is black with pale yellow lines, except for the apex of the front wing, which is entirely black; the leaflike expansions on the hind leg are mostly orange and red. **Natural history**: All members of this genus feed on passion fruit (*Passiflora*), penetrating the green flesh with their needle-like mouthparts to suck out the contents of the seeds contained within. They also feed on buds and leaves to obtain water. **Related species**: This species was previously known as *Anisoscelis flavolineata*. It belongs to the subgenus *Bitta*, as do the other 3 species of *Anisoscelis* found in Costa Rica. There are probably at least 150 species of Coreidae in Costa Rica.

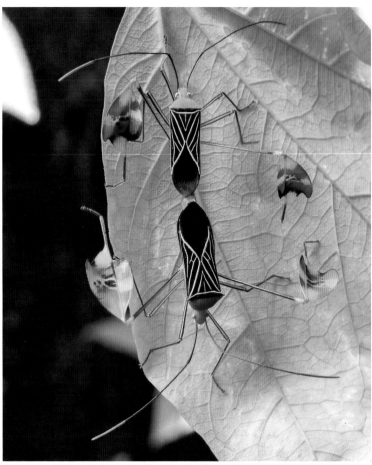

A mating pair of *Anisoscelis alipes* on *Passiflora biflora* (Passifloraceae).

Heliconia Bug (*Leptoscelis tricolor*)

Description: 1.5 cm (0.6 in) long. Although belonging to the same family as the leaf-footed bugs (Coreidae), this species does not have leaflike hind legs. Note that males do have very robust hind femora. Adults are primarily black with red lines at the edges of the thorax and abdomen; in the middle of the wings there is a pale diagonal line. Nymphs are more variable in color, apparently reflecting the background color of their host plant. **Natural history**: Both adults and nymphs use their proboscis-like mouthparts to suck sap from the flowers and fruits of heliconia plants. Males defend territories on heliconia flowers and, when an encounter between two males becomes aggressive, they turn around and grapple one another with their hind legs. After mating, females lay their eggs singly, often on several different heliconia plants. The eggs hatch in roughly two weeks; the nymphs require about a month to reach adulthood; adults can live for up to three months. **Related species**: The only other species in Costa Rica, *Leptoscelis quadrisignata*, is brownish and lives on calatheas instead of heliconias.

Leptoscelis tricolor on inflorescence of *Heliconia*.

41

Peanut-headed Bug (*Fulgora lampetis*)

Description: Up to 9 cm (3.5 in) in length. The large hollow protuberance in front of the head has false eyes and looks reptilian. **Natural history**: While peanut-headed bugs are not uncommon, they are often difficult to spot because they are well camouflaged and rest on tree trunks, often high above the ground. They feed on the sap of certain trees, including *Hymenaea courbaril* (Fabaceae), *Protium* (Burseraceae), and *Simarouba* (Simaroubaceae). When threatened they spread their wings, exposing large eyespots on the hind wings, and give off a foul odor. They are mistakenly believed to be venomous, and according to local lore, if bitten by a *machaca* (the Spanish name), a person will die unless he or she makes love within 24 hours (unfortunately they do not bite). **Related species**: There are 3 species of peanut-headed bug (genus *Fulgora*) in Costa Rica, with only subtle differences to distinguish them. They belong to the family Fulgoridae, which consists of about 30 species in Costa Rica; species in other genera often have bizarre heads, but none resemble the peanut-headed bugs.

Fulgora lampetis. The true head is located below the peanut-shaped protuberance (note the small black eye).

Wax-tailed Planthopper (*Pterodictya reticularis*)

Description: 3.5 cm (1.4 in) long. With its dark, netlike (reticulate) venation on the front wings, this planthopper somewhat resembles a cicada, but differs most notably by the large, white wax ribbons projecting from its rear end (at least in females). It also has much longer antennae. **Natural history**: Generally seen in small groups on tree trunks of *Zanthoxylum* (in the citrus family, Rutaceae), this species feeds on plant sap. The female lays an egg mass (2 cm / 0.8 in long) on a vertical surface and covers the eggs with wax ribbons from its abdomen. The nymphs also produce wax from the tip of the abdomen, but the function of this wax is not well understood. **Related species**: This is the only species in the genus in Costa Rica. It is a member of the family Fulgoridae.

Pterodictya reticularis. Note the wax ribbons at the rear end.

Emerald Cicada (*Zammara smaragdina*)

Description: 5 cm (2 in) long, including the wings. This large cicada is a bright emerald-green or blue; the front wings are transparent with spots, including a large dark spot at the apex that encloses two clear spots. The prothorax has prominent lateral flanges. Since most species inhabit the canopy, cicadas are often difficult to see in the field, but they occasionally come to lights at night. **Natural history**: Cicadas suck the watery sap from trees, which is consumed and excreted in great quantities; indeed, one can often feel a fine mist raining down from a canopy full of cicadas. In the Sarapiquí region they are particularly fond of *Pentaclethra* trees. Males sing by rapidly vibrating a pair of ribbed membranes located at the base of the abdomen. They can sing any time during the day but there are strong bursts of chorusing for 15–20 minutes at dawn and again at dusk (like many other cicada species). After mating the female inserts her eggs into saplings; when they hatch the nymphs fall to the ground and use their stout front legs to burrow into the soil, where they begin sucking sap from roots, especially roots of legume trees. The nymphs develop underground for at least a year (possibly more). Upon reaching maturity, they emerge from the soil (usually during the rainy season), crawl up the closest vertical object, and cast off their exoskeleton; the newly emerged adults then fly up into the canopy. **Related species**: There are 4 species of *Zammara* in Costa Rica and a total of 45 cicada (family Cicadidae) species. A related species, *Z. smaragdula*, occurs in the Central Valley.

Zammara smaragdina male feeding from trunk of *Vochysia* (Vochysiaceae); adult (inset) just emerged from nymphal exoskeleton.

Heliconia Spittlebug (*Mahanarva costaricensis*)

Description: 2 cm (0.8 in) in length. This species has black wings with two or three reddish-orange spots at the base of each wing and five spots in the rest of the wing. The thorax is black with a fine reddish line on each side. **Natural history**: Spittlebugs as a group (Cercopidae and a couple of related families) feed on the watery xylem sap of their host plants; the nymphs produce a frothy mass from the sap, within which they gain protection from predators and adverse physical conditions. *Mahanarva costaricensis* and *M. insignita* feed on the stems and in the floral bracts of *Heliconia* (Heliconiaceae); they are unusual among spittlebugs in that their nymphs are semi-aquatic, living in the water that accumulates inside the bracts. **Related species**: Another species also found on heliconias, *M. insignata*, is similar but has both lines and spots on the wings, and has fewer spots. The family Cercopidae consists of about 50 species in Costa Rica.

Mahanarva costaricensis adult with front legs extended in warning pose. The nymphs are hidden beneath the spittle mass.

45

Sharpshooter Leafhoppers (subfamily Cicadellinae)

This group includes some of the largest (up to 2 cm / 0.8 in) and most colorful leafhoppers. Like other leafhoppers, sharpshooters have a row of tiny spines on the hind tibia and are very good jumpers. They have a more bulging face than do other leafhoppers. The name *sharpshooter* refers to the droplets of water that are forcibly ejected from the anus, one droplet every few seconds when actively feeding. All leafhoppers suck sap from plants, but unlike other members of the family, sharpshooters specialize on xylem sap, which is mostly water. They feed on various plants, piercing small stems and large leaf veins with their needle-like mouthparts. The female inserts her eggs on the undersides of leaves, usually about a dozen in a row, which are then covered with a protective white powder secreted from the anus. There are approximately 200 species of sharpshooter in Costa Rica and at least 1000 species of leafhopper (Cicadellidae), making it the largest family of Hemiptera. Because the biology is relatively uniform among the sharpshooters, the authors present them in gallery form rather than providing accounts of individual species.

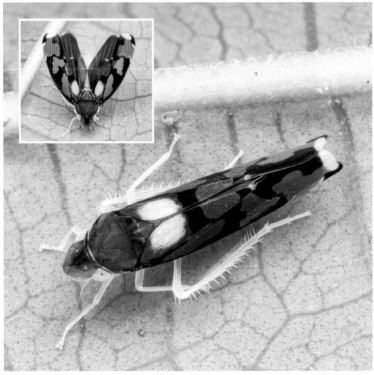

Ladoffa species. This leafhopper is often seen with its wings open (inset).

Platygonia praestantior. Found as it was feeding on inflorescence of *Heliconia.*

Paromenia species. These insects often come to lights.

Abana gigas. One of the largest leafhoppers in Costa Rica; note white waxy secretion on the surface of the body.

Oncometopia clarior on leaf of *Erythrina* (Fabaceae).

Mexican Treehopper (*Membracis mexicana*)

Description: About 1 cm (0.4 in) long. Pronotum (front of thorax) raised into a semicircular plate that is bright yellow with black spots; the wings and legs are black. The nymphs are white with black spots. **Natural history**: This species sucks sap from various plants. Males buzz their wings during courtship; females enclose their eggs with a white foamy substance. The nymphs usually feed in small groups, and ants often come to harvest the sugary liquids that these nymphs excrete. About two months are needed to go from freshly laid eggs to new adults. **Related species**: The only other species in the genus in Costa Rica is *M. dorsata*, whose adults are black and white and whose nymphs are entirely white. In total there are about 200 species of treehopper (Membracidae) in the country.

Female *Membracis mexicana*. Inset shows mature nymphs with ants (and scale insects).

Thorn Bug (*Umbonia crassicornis*)

Description: 1.5 cm (0.6 in) long. Species of *Umbonia* have a narrow, thornlike pronotum (front of thorax) that is oriented vertically; the sides and back of the pronotum are also pointed. Thorn bugs are green with reddish and yellowish stripes, although color changes as adults mature. Viewed from the side, males have a wider "thorn." **Natural history**: Thorn bugs feed on plant sap and are especially fond of leguminous trees such as *Inga* or *Calliandra*. The female inserts her eggs into living plant tissue and then stands guard over the eggs; she remains with the nymphs throughout their development. If threatened by a predator, the nymphs produce alarm calls (using vibrations transmitted via the plant surface), and the mother responds by buzzing her wings or kicking with her hind legs to drive away the intruder. Young adults generally remain with their mother and younger siblings for up to a month. When mature, the females either remain on the same plant or fly to a nearby tree, while males leave the aggregation in search of females on other plants. When a male lands on a plant, he vibrates and, if a female responds, the pair vibrate in duet until the male locates the female. **Related species**: There are 4 species of *Umbonia* reported from Costa Rica and in total about 200 species of treehopper (Membracidae).

Mother guarding her young nymphs. Note the spiral arrangement of nymphs.

Mother (far right, top of branch) with young adult females and males (with enlarged "thorn").

50

Beetles

(order Coleoptera)

On beetles the front wings are converted into hardened covers (known as elytra) that conceal and protect the hind pair of wings and abdomen; they rely solely on the hind wings for flight. The larvae inhabit rotting wood, leaf litter, living plant tissue, and even water, depending on the species. Coleoptera contains more species than any other order of animals.

Two-spotted Tiger Beetle (*Pseudoxycheila tarsalis*)

Description: 1.5 cm (0.6 in) long. Generally, tiger beetles have large eyes and long legs and can run very rapidly when approached. Note, however, that the Two-spotted has smaller eyes than most other tiger beetles. It is blue-green with a large cream-colored spot on each wing cover (elytron). **Natural history**: Tiger beetles are predators of other insects, which they capture by pouncing on and grabbing with the sharp teeth on their mandibles. If captured themselves, they secrete a defensive odor from the tip of their abdomen. This species, which is nearly flightless, hunts on relatively open ground and moist dirt banks, and is most active in the morning; males defend territories against other males. The larva lives in a narrow tunnel in the ground, where it waits for passing prey. **Related species**: This is the only species of *Pseudoxycheila* in Costa Rica, but there are nearly 60 species of tiger beetle (family Cicindelidae) in the country.

Pseudoxycheila tarsalis. Inset shows young larva in its tunnel, frontal view of head.

Bess Beetle (*Veturius sinuaticollis*)

Description: 3 cm (1.2 in) long. As a family, bess beetles are recognized by their slightly flattened, shiny black body, with a space between the thorax (pronotum) and wing covers (elytra), the latter having distinct longitudinal lines. Individual species, however, are more difficult to distinguish, at least in the field. **Natural history**: These beetles live in rotting wood. A female or male locates a suitable log (one that has been dead for at least a year), starts excavating, and is joined by a member of the opposite sex. After mating, the pair rears their offspring from larvae to adulthood. Larvae feed both on wood that has been chewed up by the parents and on the feces of adult beetles. Various microbes are present in the hind gut of adults, and some of these are excreted and continue to digest the wood in the excreted feces. In eating this microbe-rich excrement, the larvae obtain much more nitrogen than is present in the original cellulose (of the wood), and consequently they often reach maturity in just two or three months. Adults and larvae communicate via squeaking noises made by rubbing one part of their body against another. **Related species**: Costa Rica has 13 species of *Veturius* and a total of roughly 55 species of bess beetle (family Passalidae).

Veturius sinuaticollis. These beetles can be seen walking on rotting wood.

Hercules Beetle (*Dynastes septentrionalis*, previously *D. hercules*)

Description: Males 5–15 cm (2–6 in) long, including the thoracic horn; females 5–7 cm (2–2.8 in). Males have two horns, one at the front of the thorax (pronotum) and the other on the head. The thoracic horn, which has a dense covering of reddish-brown hairs on the underside, can constitute half the length of the body in large males or just one-fifth in small males. The horn on the head is slightly shorter and is gently curved upward. The thorax itself is black while the wing covers (elytra) are either black or yellowish. The females lack horns; most of the surface of the elytra, except the yellowish posterior lateral area, is rough in texture and black.

Natural history: The Hercules Beetle is nocturnal. Males use their impressive horns in fighting with one another to dominate areas on rotting tree trunks, where females are most likely to show up for mating and egg-laying. After trapping an opponent with his horns, the male pushes and shakes from side to side in an attempt to throw the other male from the log. Male size is directly related to the quality of food consumed during the larval stage, and larger males, with their disproportionately larger horns, usually win these shoving matches. After mating, the female lays her eggs in cavities in rotting trunks. Here the larvae feed and eventually transform into pupae and then adults. The entire life cycle, from egg-laying to new adult, requires about two years, and adults can live for many months. Adults are most commonly seen from April to June and rarely seen the rest of the year. **Related species**: There is only one species of *Dynastes* in Costa Rica, but about 130 species in the subfamily Dynastinae (family Scarabaeidae).

Dynastes septentrionalis. The male is more brightly colored during the day, when the air is dryer; with greater humidity at night it becomes duller.

Elephant Beetle (*Megasoma elephas*)

Description: Males 7–12 cm (2.8–4.7 in) long, including the horn; females 5–8 cm (2–3.1 in). The male has a long, upward curving horn on its head; the tip bifurcates into two points, the base has a broad protuberance above. In large males, this horn can be as long as the width of the body. The front corners of the thorax (pronotum) have sharply pointed projections. Nearly the entire body is densely covered with tiny, brownish hairs. The front legs of the males are noticeably longer than the other legs. The female lacks a horn, and only the middle and back of the body have brownish hairs. **Natural history**: Mostly nocturnal. Large numbers of adults (in some cases more than 30) have been seen feeding on the bark and sap of certain trees (*Lonchocarpus*, *Tabebuia*, *Handroanthus*, and *Delonix regia*). Males fight for territories on tree branches, where females are likely to show up for mating and feeding. Male size is directly related to the quality of food consumed during the larval stage; larger males, with their disproportionately larger horns, usually dominate in the competition among males. Larvae feed in the interior of standing, rotting tree trunks. The complete life cycle from laying eggs to the emergence of new adults probably requires two years. Adults can be seen throughout the year but are most abundant in May. **Related species**: There is only one species of *Megasoma* in Costa Rica, but about 130 species in the subfamily Dynastinae (family Scarabaeidae).

Megasoma elephas. Both sexes have velvetlike brownish hairs.

Jewel Scarab (*Chrysina chrysargyrea*)

Description: Up to 3 cm (1.2 in) long. A refulgent silver overall, occasionally gold or orange-gold, with copper-brown on the front half of the head, the sides of the thorax (pronotum), and most of the legs. Also note the metallic blue on the tips of the legs (the tarsi). **Natural history**: The females lay their eggs in fallen tree trunks on the forest floor; the larvae feed on rotting wood. The latter pass through three larval stages, after which they construct a chamber in which to pupate, then emerge as adults. The complete life cycle requires about one year. Adults live and feed on foliage in the canopy; because they live high up, are inactive during the day, and have a body that reflects the colors of their surroundings, they are difficult to observe. **Related species**: In Costa Rica, the genus *Chrysina* comprises 26 species; these include some of the most brilliantly colored beetles in the country, varying from light shiny green to metallic silver or gold. The genus belongs to a group known as the "shining leaf chafers" (subfamily Rutelinae: family Scarabaeidae), which consists of around 235 species in Costa Rica.

Chrysina chrysargyrea; a silver male mating with a golden female.

Giant Metallic Ceiba Borer (*Euchroma gigantea*)

Description: 5–6 cm (2–2.4 in) long; the largest member of the family in tropical America. The wing covers (elytra) are golden-green with reddish hues in the center; the thorax (pronotum) has two large dark spots, one on each side. Recently emerged adults are covered with a yellowish, waxy powder that covers the brilliant metallic colors, but this powder falls off over time. **Natural history**: These beetles can occasionally be seen flying in open areas or resting on foliage but are most common on the trunks of living kapok (ceiba), balsa, and related trees. They are especially attracted to the fallen trunks of these trees, where adult beetles interact with one another and females lay eggs. The larvae are wood borers in fallen tree trunks and, along with other insects, contribute to their decomposition. **Related species**: There is only one species of *Euchroma* in Costa Rica but well over 500 species of metallic wood-boring beetles (family Buprestidae); most of them are small.

Euchroma gigantea. Note yellow powder on the elytra.

Agrilus species (1.2 cm / 0.5 in).

Chrysobothris delectabilis (1 cm / 0.4 in).

57

Semiotus Click Beetle (*Semiotus illigeri*)

Description: 3 cm (1.2 in) long. Click beetles (family Elateridae) can snap upward into the air when placed on their back. They have a long, shiny body and their large prothorax has pointed hind corners. Click beetles share these characteristics with false click beetles (Eucnemidae), but the latter have a more rounded head. The Semiotus Click Beetle can be recognized by the three spines on the front of the head (the middle one is quite small) and the pattern of yellow and black stripes on the wing covers (elytra); the black mark on the prothorax is variable. **Natural history**: The click mechanism probably serves as a defense against predators, since a beetle that snaps back and forth is more difficult to grasp. Adult click beetles feed on pollen, nectar, rotting fruit, and soft-bodied insects. The larvae are elongate and have cuticle that is tougher than that of most beetle grubs, hence the common name *wireworm*. Larvae of this genus live in rotting wood, where they feed on fly larvae. **Related species**: There are at least a dozen species of *Semiotus* in Costa Rica, and a total of 400–500 species of click beetle (Elateridae) in the country.

with wings open

Semiotus illigeri. This is one of the more striking click beetles; it is active during the day.

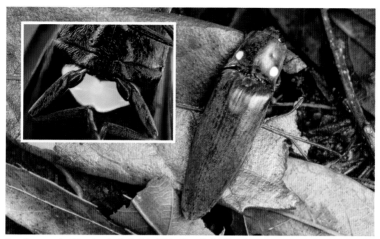

Pyrophorus species. Though not a firefly, this click beetle emits light from two spots on the pronotum and a patch on the ventral surface (inset); the latter is only visible when it flies.

Firefly (*Aspisoma* species)

map for genus

Description: 1.5 cm (0.6 in) long. Body is oval-shaped when viewed from above, and not as hard as in other beetles. The front part of the thorax (pronotum) is rounded in front and completely covers the head. Wing covers (elytra) pale yellow with wide, dark brown stripes in the center; thorax (pronotum) with a pair of reddish marks. **Natural history**: Males produce flashing, yellowish-green lights during their slow flight shortly after sunset, and if a female in the vegetation is interested she responds with a brief, weaker flash. Adults can also be found on the upper

Aspisoma species

parts of bushes before sunset. As far as we know, the adults do not eat, though it is possible that they imbibe some nectar. Like other fireflies, the larvae have plate-like segments over the top of the body and long piercing mandibles for injecting enzymes into their prey; in the case of *Aspisoma*, the prey consists of snails. The larvae have a pair of organs at their rear end that emit a sporadic light that lasts for about 5 seconds at a time, but note that the entire body casts a faint glow. This probably serves to warn potential predators of their chemical defenses, since both larvae and adults produce toxic steroids (similar to those in toads). **Related species**: There are at least 6 species of *Aspisoma* in Costa Rica and perhaps around 250 species of firefly (family Lampyridae).

Photuris crassa, another species of firefly.

A mating pair of *Photinus* fireflies on the underside of a leaf.

Net-winged Beetle (*Caenia kirschi*)

Description: 1.5 cm (0.6 in) long. This beetle's body is not as hard as in most other beetles and it is widest near the rear end. Like other members of its family, on this species the wing covers (elytra) are crisscrossed by ridges, forming a netlike pattern. It is bright red overall, with an orangish thorax (pronotum) and black legs. The pronotum is triangular in front; the back corners are pointed. In both sexes, note the branched antenna. **Natural history**: These beetles harbor distasteful chemicals, which they advertise with their bright colors. Because they repel predators, net-winged beetles serve as models for other insects that mimic them. Adults occur on flowers and leaves, where they feed on nectar and honeydew. The larvae inhabit rotting wood; their feeding habits are still poorly known. **Related species**: There are at least 3 species of *Caenia* in Costa Rica and about 200 species of net-winged beetles (family Lycidae).

Caenia kirschi. The netlike pattern on the elytra is a characteristic of the entire family.

61

Soldier Beetle (*Chauliognathus heros*)

Description: 1.5–2.0 cm (0.6–0.8 in) long. Beetles in the family Cantharidae have a softer body than most other beetles and are sometimes confused with the closely related fireflies. Unlike the latter, soldier beetles have their head completely exposed (not concealed to some degree by the thorax). The thorax (pronotum) is yellow with a central black spot, and the wing covers (elytra) are yellow in front and black behind. **Natural history**: Soldier beetles generally move rather slowly, probably because they are protected by defensive chemical compounds secreted from glands located throughout their body; these compounds have been shown to deter jumping spiders. Adults feed on nectar from flowers; if the supply of nectar is scarce, individual beetles will defend a group of flowers to give time for more nectar to be secreted. The larvae have a velvety texture. They can sometimes be seen walking over the ground or in humid leaf litter; they feed on small invertebrates. **Related species**: About 20 species of *Chauliognathus* are currently known from Costa Rica and there are an estimated 400 species of Cantharidae in the country.

Chauliognathus heros

Ironclad Beetle (*Zopherus jansoni*)

Description: 2–3 cm (0.8–1.2 in) in length (males are smaller than females). Body peanut-shaped when viewed from above, with a bumpy surface. Ironclad beetles derive their name from their extremely hard body. **Natural history**: This species is usually found on tree trunks, where it feeds on fungi. During the day, it usually hides on moss-covered trunks; if disturbed it falls to the ground and plays dead. The wing covers are fused, and so these beetles cannot fly. The larvae live in tunnels in decomposing wood and feed on certain types of fungi. Larvae and adults can live for several years, which is quite a long time for an insect. **Related species**: There are 4 species of ironclad beetle (Zopheridae: tribe Zopherini) in Costa Rica, 3 of them in the genus *Zopherus*.

Zopherus jansoni. Ironclad beetles can survive being run over by a car. By studying the structure of these beetles, scientists hope to gain inspiration for developing tough, yet flexible materials.

63

Harlequin Longhorn Beetle (*Acrocinus longimanus*)

Description: 7.5 cm (3 in) in length. Body with elaborate patterns of black, reddish, and paler colors. Males have extremely long front legs. **Natural history**: Despite the gaudy colors, the Harlequin Longhorn is fairly well concealed when it rests on lichen-covered bark. Adults feed on tree sap. The female lays her eggs in recently dead or unhealthy trees, such as those infected with fungi; she begins by gnawing a hole in the bark and lays about 20 eggs over the course of three weeks. The larvae bore into the wood and feed for 7 to 8 months and then spend another 4 months in the pupal stage. **Related species**: There are more than 1100 species of longhorn beetle (family Cerambycidae) in Costa Rica. Nearly all of them have long narrow bodies with very long antennae (hence the name *longhorn*), but no other species resembles the Harlequin Longhorn.

Female *Acrocinus longimanus*

Harlequin Longhorn Beetles often carry pseudoscorpions (in circle) and mites (tiny red dots). Lacking wings, they use the harlequin beetles as a means of transport to another patch of dead wood, but do no harm to their beetle carrier.

Chlorida cincta is one of the more common longhorn beetles in cloud forests, where they are usually seen at night.

Parandra species, unlike most longhorn beetles, have short antennae.

These leaf beetles have a very robust body and a relatively wide head. Adults and larvae feed on the foliage of their host plant. The larvae of some species synthesize chemical defenses in specialized glands, while those of other species sequester compounds from plants and use them as chemical defenses.

Inkblot Beetle (*Calligrapha fulvipes*)

Description: 1 cm (0.4 in) long. The wing covers (elytra) are light lime-green with elaborate black markings that can vary from individual to individual. The thorax (pronotum) is dark with a metallic sheen and the legs are reddish-brown. **Natural history**: Adults have been observed on certain plants in the mallow family (Malvaceae: *Sida*) and it is likely that the larvae also feed on these plants. Larvae of other species in this genus are leaf skeletonizers, consuming the entire leaf except for the veins; when young they feed in groups. A closely related species (*C. pantherina*) was released in Australia as a biological control agent of *Sida* (these plants are invasive weeds in Australia). A few, more distantly related North American species are of biological interest because they are asexual (females reproducing via parthenogenesis, without males). **Related species**: There are at least 16 species of *Calligrapha* in Costa Rica

Calligrapha fulvipes on *Sida* (Malvaceae).

Zygogramma violaceomaculata on *Malvaviscus* (Malvaceae). This species in similar in appearance yet belongs to a different genus.

Black-bellied Leptinotarsa (*Leptinotarsa undecimlineata*)

Description: 1 cm (0.4 in) long. The wing covers (elytra) are whitish to pale greenish-gray, with nine longitudinal black stripes. The thorax (pronotum) is yellow with irregular black markings. The legs and antennae are black. **Natural history**: Adults and larvae feed on the leaves of *Solanum lanceolatum* (which is in the same genus as the potato). The plump larvae are white with black markings. **Related species**: There are 7 species of *Leptinotarsa* in Costa Rica. They belong to the subfamily Chrysomelinae.

Leptinotarsa undecimlineata, a close relative of the infamous Colorado Potato Beetle (*L. decimilineata*). Also note larvae (inset).

Tortoise Beetles (Chrysomelidae: subfamily Cassidinae)

On many species in this subfamily, the thorax and wing covers (elytra) are expanded on the sides, and thus cover the legs and head, hence the name *tortoise beetle*. All of them have a very small mouth. Adults and larvae feed on the foliage of their host plant.

Palm-skeletonizing Beetle (*Alurnus ornatus*)

Description: 2–2.5 cm (0.8–1 in) long. This is one of the largest of the tortoise beetles—and the least tortoise-like. The wing covers (elytra) are pale yellow with black spots; the thorax (pronotum) is reddish. **Natural history**: Adults and larvae feed on palms (Chamaedorea); the larvae skeletonize tender leaves. The flattened, oval-shaped eggs are laid in short chains and hatch in a month or so. The larvae need about 8 months until they are ready for pupation; the pupal stage lasts another month and adults live for 2–6 months. **Related species**: There is only one other species of *Alurnus* in Costa Rica, *A. salvini*, which has a black pronotum. These two are the only representatives of the tribe Alurnini in the country.

A mating pair of *Alurnus ornatus*. This species can be seen flying from one palm tree to another.

Golden Tortoise Beetle (*Charidotella egregia*)

Description: 1 cm (0.4 in) long. This beetle is circular when viewed from above, slightly dome-shaped in lateral view; the head is not visible from above. Normally metallic gold, this species turns red when disturbed. The normal gold color is due to reflection, which is in part a consequence of a liquid that fills numerous microscopic channels. When the beetle is disturbed it withdraws this liquid and the wing covers (elytra) then become translucent, thereby providing an unobstructed view of an underlying reddish pigment. **Natural history**: Adults and larvae feed on the leaves of plants in the morning glory family (Convolvulaceae). The larvae use movable appendages (caudal processes) on the tip of the abdomen to place molted skins onto their back and also use their eversible (telescoping) rectum to add excrement to the skins; the odor from this covering provides the larva with protection against predators. **Related species**: There are about 20 species of *Charidotella* in Costa Rica; the genus belongs to the tribe Cassidini.

Charidotella egregia, with normal gold coloration.

This individual is in the process of turning red.

69

Golden Target Beetle (*Ischnocodia annulus*)

Description: 0.5–1 cm (0.2–0.4 in) long. Also known as the Ringed Tortoise Beetle; both names refer to the alternating black and yellow rings. When seen from above, the outer margins of the wing covers (elytra) and thorax (pronotum) are essentially transparent; thus, although the thorax completely covers the head, it is still visible from above. **Natural history**: Adults and larvae feed on the leaves of *Cordia* (Boraginaceae). **Related species**: This is the only species in the genus in Costa Rica; it belongs to the tribe Cassidini.

Ischnocodia annulus feeding on Cordia leaf.

Omaspides Tortoise Beetle (*Omaspides convexicollis*)

Description: 1.5 cm (0.6 in) "long," though note that the body is about as wide as it is long. Shiny black, with a large spot on each wing cover (elytron). The back corners of the thorax (pronotum) are pointed. **Natural history**: Adults and larvae feed on leaves of plants in the morning glory family (Convolvulaceae). Eggs are laid in a group, with each egg attached to the underside of the leaf by a thread. The female stands guard over both the egg mass and the larvae that hatch from the eggs; she does not feed while guarding the eggs. The time required to go from freshly laid egg to adult is nearly two months. **Related species**: There is only one other species of *Omaspides* in Costa Rica, *O. bistriata*; the genus belongs to the tribe Mesomphalini.

The male *Omaspides convexicollis* is darker than the female and has large orange spots.

Mother protecting her eggs from parasitic wasps that are crawling over her back.

71

Weevils (Curculionidae and Dryophthoridae)

Weevils are recognized by their long snout, which is actually a prolongation of the face; their mandibles are located at the tip of the snout. The only other beetles with such a snout are members of the same superfamily (Curculionoidea), including, for example, the straight-snouted weevils (Brentidae). A few weevils have a very reduced snout, most notably the bark beetles (subfamily Scolytinae). Females of most species use their snout and mandibles to drill into and pry open tough plant tissue. After penetrating the plant, the female turns around, deposits an egg in the hole, and then seals it up with a cementlike secretion. Weevil larvae lack legs, and the larvae of most species spend their lives burrowing in and eating plant tissue. Costa Rica has an estimated 6000 species of weevil.

Bearded Palm Weevil (*Rhinostomus barbirostris*)

Description: Up to 5 cm (2 in) long; including the long snout, this is one of the largest weevils in the world. The body is black and covered with tiny punctures, with longitudinal lines on the wing covers. Males have much longer front legs and a longer snout, which is covered with reddish-brown hairs. **Natural history**: The female uses the mandibles at the tip of her snout to drill a hole into palm trunks, then turns around to lay an egg in the hole, and finally plugs the hole with a secretion from the tip of her abdomen. She then walks a small distance and repeats the process. Males move rapidly across the surface of trunks searching for females in the process of drilling; upon encountering such a female, he wipes her back with the hairs on his snout by moving his snout from side to side. He then attempts to mate with the female, and if she accepts, they mate for about a minute. If another male approaches, the males often fight by hitting one another with their snouts in an attempt to knock the opponent from the trunk. The larvae bore tunnels and feed inside the palm trunk; the cycle from egg to adult takes 4–6 months. **Related species**: In Costa Rica, the only other species in the genus is the very similar *R. thompsoni*. They belong to the family Dryophthoridae, which comprises 130 species in the country.

Male *Rhinostomus barbirostris*. The plump larvae of this species and another palm weevil (*Rhynchophorus palmarum*) are prized as food by indigenous people of the Amazon.

Jekel's Broad-nosed Weevil (*Exophthalmus jekelianus*)

Description: 1–1.5 cm (0.4–0.6 in) long. Broad-nosed weevils (subfamily Entiminae) have a short, wide snout. In this species the body is black underneath and covered with tiny iridescent scales of various colors, primarily blue-green but also light blue and yellowish-green. **Natural history**: Adults feed on the leaves, buds, and flowers of various plants and can be pests of coffee, especially in shaded habitats. When they feed, they make irregular cuts on the edges of leaves. Unlike most weevils, which use their snout to drill an egg hole in plant tissue, females of broad-nosed weevils lay their eggs in clusters on the surface of leaves. The hatching larvae drop to the ground and feed on roots. Pupation takes place in an underground chamber. **Related species**: There are about 25 species of *Exophthalmus* in Costa Rica; the genus belongs to the subfamily Entiminae.

Exophthalmus jekelianus

Exophthalmus nicaraguensis, one of the larger (2 cm / 0.8 in) species in the genus.

73

Splendid Weevil (*Eurhinus magnificus*)

Description: 0.5 cm (0.2 in) long. Bright metallic green, but with some reddish-copper reflections, especially on the rostrum and legs. **Natural history**: Like other members of the genus, *E. magnificus* is associated with species of *Cissus*, which are vines in the grape family (Vitaceae). Adults feed mostly on the stems and females lay eggs in young parts of the stem by creating a cavity with their rostrum and depositing an egg. Larva feed inside the stem, causing it to become swollen (gall formation); when mature, the galls measure about 1 cm (0.4 in) in diameter. **Related species**: There are 5 species in this genus in Costa Rica; they belong to the subfamily Baridinae.

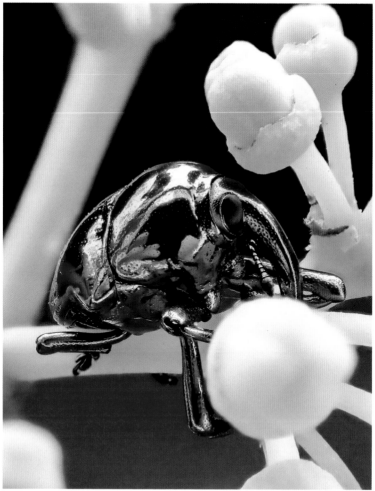

Eurhinus magnificus visiting flowers on its host plant, a *Cissus* vine.

Black and White Weevil (*Peridinetus cretaceus*)

Description: 1 cm (0.4 in) long. Black with a large white patch on the side of the thorax and two large white patches on the sides of the wing cover (elytron). **Natural history**: This genus, along with several related genera (*Ambates, Embates, Pantoteles*), is associated with black-pepper bushes (*Piper*). Adults have the curious habit of resting sideways inside a hole that they cut into leaves, which makes them look like a fallen flower or bird dropping. The larvae bore into the stems of black-pepper plants. **Related species**: There are at least 20 species of this genus in Costa Rica; they belong to the subfamily Baridinae.

Mating pair of *Peridinetus cretaceus*

Cholus Weevil (*Cholus costaricensis*)

Description: 2 cm (0.8 in) long. This species is black with two white transverse stripes. **Natural history**: Adults have been seen on *Heliconia* flowers but very little is known about their biology. **Related species**: In Costa Rica there are probably around 60 species of *Cholus*, many of them without names; the genus belongs to the subfamily Molytinae.

Cholus costaricensis

This distinctly different species, *Cholus viduatus*, was found feeding on *Vernonia patens* (Asteraceae).

Wasps, Bees, and Ants

(order Hymenoptera)

All members of this order share an interesting biological characteristic: instead of x and y chromosomes, females are produced from fertilized eggs, while males are produced from unfertilized eggs. The vast majority of species are parasitic wasps (e.g., ichneumon wasps). This order includes more species of eusocial insect (colonies with queens and workers) than any other group: paper wasps, a few bees (e.g., stingless bees), and all ants. Termites are basically the only other eusocial insects. It should be noted that among the eusocial hymenopterans all workers are female; the main difference between workers and queens is that the larvae of the latter receive more food.

Ichneumon Wasps (Ichneumonidae)

Species in this large and diverse family vary tremendously in size and color. Their diagnostic characteristics may be difficult to observe in the field, but note antenna that almost always have 16 or more segments and front wings with a large, curved cell resembling the head of a dog. Adults feed on nectar while the larvae parasitize other insects or spiders; the female lays an egg on or in the host and the larva slowly consumes it, eventually killing it. The females of some species inject a venom that paralyzes the host; in other species, the host remains active as the "alien" eats it. The most common insect hosts are larvae of moths, butterflies, beetles, and flies (each ichneumon species parasitizes only a restricted set of hosts). This family includes more than 3000 species in Costa Rica, as does the related family of parasitic wasps, Braconidae. Below is a selection of interesting species that belong to the family.

Epirhyssa mexicana. This female is drilling into wood with her ovipositor in an effort to paralyze a beetle larva and lay an egg on it.

Messatoporus species (possibly *M. longitergus*). This female is ovipositing into the nest of a mud dauber (*Sceliphron*) wasp; the ichneumon larva parasitizes the mud dauber larva.

Neotheronia species; female in flight (inset). Their wings emit a blue glow, especially while flying.

Pimpla ichneumoniformis. This species appears to mimic the stinging paper wasp, *Agelaia panamensis.* Its larvae parasitize moth pupae.

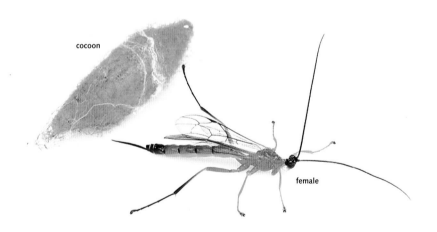

cocoon

female

Hymenoepimecis robertsae. The larva of this species feeds on (and eventually kills) the Golden Silk Orbweaver (p. 181) while attached to the outside of the orbweaver's body.

Netelia. Wasps in this genus search for caterpillars at night. Their larvae feed on a caterpillar while attached to the outside of its body, and eventually kill it.

Thyreodon laticinctus prefers the shaded understory of rainforests. The larva of this wasp parasitizes the larvae of hawk moths (Sphingidae).

Fig Wasp (*Pegoscapus silvestrii*)

Description: 0.2 cm (0.1 in) long. Though abundant, these wasps often go unnoticed due to their minute size. The best way to see them is to split open a nearly ripe fig. The females have a flattened head and the males are wingless. Each species of fig tree (Moraceae: *Ficus*) has its own species of fig wasp, so the easiest way to identify the wasp is to identify the fig tree from which it came. *Pegoscapus silvestrii* is associated with *Ficus pertusa*. **Natural history**: Fig wasps are the exclusive pollinators of fig trees, whose minute flowers are concealed inside the fig. When the female flowers are ready to be pollinated, the tree gives off an odor that attracts a female fig wasp, who enter the green fruit through an opening so narrow that her antennae and wings often break off as she squeezes through this hole. Once inside, she pollinates the flowers and then lays eggs in the fig ovules that are closest to the central cavity. Ovules that receive eggs produce fig wasps instead of seeds, so the tree loses some of its seeds but gains the services of a very reliable pollinator. The tree often aborts green fruits that haven't been pollinated and thus by pollinating the flowers the female wasp ensures that her offspring will not be aborted. Once the female wasp has laid her eggs, she dies inside the fruit. Each fig wasp larva feeds inside a fig ovule for a few months, pupates, and emerges as an adult into the inner cavity of the fruit. The blind, wingless males emerge first and, after mating, they chew an exit hole through which the females depart for the outside world; the males, however, shall never experience the outside world since they die inside the fruit (or fall to the ground and die). At this point, just before the figs are fully ripe, the masculine flowers have matured and the female wasps, before leaving the fruit, pack pollen into special pockets in their bodies. When female fig wasps depart the tree in which they were born, they must locate another tree of the same species that is in just the right stage to receive them. They live for less than a week but are thought to be capable of dispersing over 50 km (30 miles) when assisted by wind currents. **Related species**: In Costa Rica there are about 40 species of fig-pollinating wasp. Most belong to the genus *Pegoscapus* and are associated with strangler figs; other species belong to the genus *Tetrapus* and are associated with fig trees that are not strangler figs. All belong to the family Agaonidae.

Female *Pegoscapus silvestrii* recently emerged from this nearly ripe fig. Figs of *Ficus pertusa* (inset).

Velvet Ant (*Pseudomethoca chontalensis*)

Description: 1 cm (0.4 in) long. Velvet ants are not in fact ants, although the females do resemble ants (males have wings and resemble wasps). This genus is characterized by its round (as opposed to oval) eyes and females that have a square-shaped head when viewed from above. **Natural history**: Though not aggressive, the female has a very painful sting, a defensive weapon that she advertises with her bright colors. She can thus afford to calmly walk about the forest while searching for nests of solitary bees. When she locates a nest containing mature bee larvae or pupae, she chews a hole with saliva and debris. Upon hatching, the larval velvet ant (which is an ectoparasitoid) slowly feeds on the immature bee, eventually killing it. **Related species**: There are at least 40 species of *Pseudomethoca* in Costa Rica and in total probably as many as 300 species of velvet ant (family Mutillidae).

Female *Pseudomethoca* species, possibly *P. chontalensis*.

Dormant aggregation of female *Pseudomethoca* inside a hollow twig on the ground.

Tarantula Hawk (*Pepsis aquila*)

map for genus

Description: Up to 4 cm (1.6 in) long, excluding the antennae and wings. This is one of the largest wasps in the country (another *Pepsis* species slightly eclipses this one, reaching 5 cm / 2 in). The body and legs are black with metallic blue reflections; the front wings are reddish-orange with a dark band at the extreme apex; the antennae are mostly pale yellow. Like other spider wasps (family Pompilidae), tarantula wasps periodically flick their wings as they walk about (the reason for this is unclear). **Natural history**: Adults of this species are frequent visitors and potential pollinators of certain species of passion flower (*Passiflora*). They are not at all aggressive toward humans, but if handled the female can deliver a very painful sting. The female wasp searches for tarantulas in their burrows and, on encountering one, quickly stings it with a paralyzing venom, lays an egg on the surface of the body, then covers the entrance to the burrow. Upon hatching, the larva of the wasp feeds on the paralyzed tarantula, eventually killing it and consuming nearly the entire body; it then pupates and later emerges as an adult wasp. **Related species**: There are at least 25 species of tarantula hawk (*Pepsis*) in Costa Rica; of these, 10 have orange wings (the others have black wings). Note that a few species of *Hemipepsis* have similar orange wings with a black body but differ in their wing venation. These two genera and many others belong to the spider wasp family (Pompilidae), which consists of around 250 species in the country.

Pepsis aquila walking in open field.

Mischocyttarus Paper Wasp (*Mischocyttarus basimacula*)

Description: 1.5 cm (0.6 in) long. Members of this genus have asymmetrical tarsal segments, a feature that is difficult to see without magnification. An easier way to identify them is by their small nests, which have exposed cells placed on the undersides of leaves and other protected sites. Members of the genus *Polistes* have similar nests, but these wasps are larger (more than 2 cm / 0.8 in). **Natural history**: Like all paper wasps, this species lives in colonies and builds a nest from masticated wood fibers. *Mischocyttarus* and *Polistes* generally have just one egg-laying queen at a time; she can only be distinguished by her behavior (i.e., she is the only wasp that nevers leaves the nest). The other wasps are workers that bring back insect prey with which to feed the larvae. Like most *Mischocyttarus*, this species is usually not aggressive, and its nest can generally be approached very closely without provoking them; however, if you grab one, it can sting. New colonies are initiated by a single queen who may eventually be joined by one or two "cofoundresses" (potential queens). **Related species**: There are at least 20 species of *Mischocyttarus* in Costa Rica and a total of more than 100 species of paper wasp (Vespidae; subfamily Polistinae). Several other species of *Mischocyttarus* have the same yellow and black color pattern; *M. costaricensis*, for example, is very similar but has 3 yellow spots on the upper face, whereas *M. basimacula* has just one large yellow spot.

Mischocyttarus basimacula. Eggs visible in cells at the top of the nest.

M. tolensis, with a completely different color. Eggs visible in cells at the top of the nest.

Polybia Paper Wasp (*Polybia emaciata*)

Description: 1.5 cm (0.6 in) long. To identify this species look for the combination of two things—the yellow and black coloring of the wasp itself and the structure of its nest. While many other paper wasps are yellow and black, few have a nest that is pear-shaped and built from mud (the vast majority of paper wasps build their nests primarily of masticated wood and fibers). **Natural history**: Like other paper wasps, this species lives in colonies, but unlike *Mischocyttarus* (p. 85) the colony contains numerous queens. While there is no obvious physical difference between queens and workers, the two castes differ in their behavior. Queens do virtually nothing other than lay eggs, whereas the workers search for food (insect prey and nectar) and care for the larvae. This is one of the least aggressive species of *Polybia*, but they will of course sting if their nest is threatened. New colonies are initiated when several workers and queens depart the maternal nest. **Related species**: There are 19 species of *Polybia* in Costa Rica; the genus belongs to the subfamily Polistinae and the family Vespidae.

Polybia emaciata workers on their nest; close-up of the worker (inset).

Jewel Wasp (*Ampulex* species)

map for genus

Description: 2–3 cm (0.8–1.2 in) long. Bright metallic green, with a long, narrow body. Jewel Wasps are generally seen walking over the ground while moving their antennae up and down. **Natural history**: The female spends most of her adult life searching for cockroaches. On encountering a suitable cockroach, she quickly pounces on it and uses her stinger to inject venom into the central nervous system of her victim, causing a temporary paralysis of the front legs. Unable to move for the time being, the cockroach spends half an hour grooming itself while the wasp searches for a suitable nesting site. When the wasp returns to its prey, it cuts one of the cockroach antennae in half, ingests the blood that oozes from the cut, and then grabs the cockroach by this antenna and leads it toward the nest, walking backwards. This unique behavior is only possible because the venom converts the cockroach into an obedient zombie, capable of walking but now lacking its normal escape instinct. In the nest (almost any nearby cavity), the wasp lays an egg on the underside of the cockroach and then plugs the nest entrance with debris. In about two days the egg hatches and the larva begins feeding on the living cockroach, first from the outside, and then burrowing into its abdomen. After about 40 days the larva transforms into an adult wasp. **Related species**: There are at least 6 species of *Ampulex* in Costa Rica; note that one species is black rather than a metallic color. Species in the other two genera of Ampulicidae occurring in the country are also black.

Female *Ampulex* species searching for a cockroach.

87

Horse Guard Wasp (*Stictia signata*)

Description: 2.5 cm (1 in) long. Like other sand wasps (3 subtribes of the tribe Bembicini), this species is a yellow and black wasp that can often be seen nesting on sandy beaches. **Natural history**: The female hunts for various types of flies, including deer flies resting on or flying near large mammals (even horses). She uses her front and middle legs to grasp the fly while stinging it with a paralyzing venom. Prior to hunting, the female digs a nest in suitable sandy soil, often in sites where other females are nesting. Each nest is a simple burrow ending in a single chamber; the female lays an egg on the first paralyzed fly she brings into the chamber and then leaves to hunt for more flies; but before leaving, she places a temporary plug over the entrance to discourage unwanted intruders. When there are enough flies for the larva to complete its development she permanently plugs the entrance and begins a new nest. **Related species**: In Costa Rica there are 4 species of *Stictia*; the tribe Bembicini belongs to the family Crabronidae, which probably comprises around 300 species in the country.

Female at her nest in the sand. There is no need to fear this wasp, which you will sometimes see at the beach.

Organ Pipe Mud Dauber (*Trypoxylon monteverdeae*)

Description: 1.5 cm (0.6 in) long. Most *Trypoxylon* species have an elongate, black body. Few species, however, combine these characteristics: a large size; black body with white tarsi; and nests made of vertical mud tubes open at the bottom and placed side by side (hence the common name). **Natural history**: All species in the genus *Trypoxylon* feed their larvae with spiders. This species and others in the subgenus (*Trypargilum*) are very unusual among solitary nesting wasps in that males guard the nest while the female is out hunting (in the clear majority of species, the male disappears after mating). The male even assists the females in nest construction by smoothing the wet inside walls of the nest. **Related species**: *Trypoxylon* is the largest genus of Crabronidae in Costa Rica, with about 80 species (many unnamed). Only about 6 construct organ-pipe nests (species in the *albitarse* group of the subgenus *Trypargilum*).

Organ Pipe Mud Dauber nest with outer layer of mud removed, showing cells arranged in linear series; above the adult wasp is a cell with paralyzed spiders (the larval food). Note the "organ pipe" shape of the nest (inset).

Orchid Bees (*Euglossa* species)

map for genus

Description: 1.1–1.5 cm (0.4–0.6 in) long. Members of the genus *Euglossa* display shiny metallic green, bronze, or blue, depending on the species. Orchid bees in general (Apidae: tribe Euglossini) can be recognized by their very long tongue, which is often about as long as the body; on males, the tibia of the hind leg is greatly swollen, while females have a flat, wide hind tibia for carrying pollen (this latter trait is shared with bumble bees, stingless bees, and honey bees). **Natural history**: The female builds a multi-chambered nest that she provisions with pollen and nectar as food for the larva. After provisioning one of the chambers, she lays an egg in it, closes it, builds another chamber, and repeats the process. Nests are built in a sheltered location and consist largely of plant resins. In some species, 2 to 5 females may nest together, though each female has her own set of chambers. Male bees in general do not visit flowers as frequently as females, since they do not assist in provisioning the nest; they require just a little nectar for their own energy requirements. Male orchid bees, however, are unique in that they visit certain types of orchids, as well as certain other flowers, to gather fragrances that are produced by the flowers. These fragrances are stored in the swollen hind legs, and during courtship the male hovers above a prospective mate while dispersing perfume. About 10% of orchids in tropical America depend on male orchid bees for their pollination. **Related species**: There are about 70 species of orchid bee in the country, of which 40 belong to the genus *Euglossa*. The other genera include *Eufriesea*, *Eulaema*, and *Exaerete*; the last is quite similar to *Euglossa* but larger and not commonly seen; the first two are hairy and might be confused with a bumblebee.

Male bees visiting flowers of *Anthurium* (Araceae). The bee in the center is carrying three pollinaria (pollen packets) from a *Gongora* orchid.

Female *Euglossa* bee approaching flowers of *Stachytarpheta frantzii* (Verbenaceae). Unlike males, which have the hind leg (tibia) swollen, females have a flattened hind tibia for carrying pollen. Both sexes of orchid bees have a long tongue.

Mariola Bee (*Tetragonisca angustula*)

Description: 0.5 cm (0.2 in) long. As a group, stingless bees (Apidae: tribe Meliponini) can be recognized by their wide hind legs (tibia) used for carrying pollen (in flight, the legs hang downward). They are generally smaller than honey bees. The majority are black, but the Mariola is yellowish or orangish; this species is also thinner than most other stingless bees. **Natural history**: As the name suggests, stingless bees lack a stinger, although a few species bite if you approach their nest too closely; the Mariola, however, is usually quite docile. This species is so common and widespread, even in cities, that Costa Ricans have given it several other endearing names, including *mariola*, *mariaseca*, *mariquita*, and *angelita*. All stingless bees live in colonies containing a single queen and hundreds to thousands of worker bees. The latter gather nectar and pollen from a wide diversity of flowers and are important pollinators. Most species nest in hollow trees, with only the entrance being visible, and in the case of the Mariola the entrance is a delicate tube made of light yellowish wax, 4–7 cm (1.6–2.8 in) long and about 1 cm (0.4 in) in diameter. Many stingless bees have guards sitting in the entrance tube, but the Mariola is unusual in that it also has guards hovering in front of the nest. One of the main functions of these guard bees is to deter robber bees (stingless bees belonging to the genus *Lestrimelitta*) from entering the nest and stealing food. **Related species**: There are more than 50 species of stingless bee in Costa Rica.

Tiny Mariola Bees approaching the entrance to their nest.

Bullet Ant (*Paraponera clavata*)

Description: 2–3 cm (0.7–1.2 in) in length. This is the largest ant in Costa Rica. The queen, which is seldom seen, is similar to the worker ants but begins her adult life with wings. **Natural history**: Bullet ants usually nest in the ground at the base of a tree. The colonies are smaller than those of many other ants, usually consisting of no more than several hundred workers. The latter forage in trees, hunting small insects and gathering substantial quantities of sugary liquids, which they carry back to the nest as droplets suspended from opened mandibles. These ants are famous for their very painful sting, and the intense throbbing pain can last for 24 hours or more. Visitors to the lowland rainforests of Costa Rica are advised to avoid grabbing on to trees, or to only do so with caution. **Related species**: The bullet ant is the only member of the subfamily Paraponerinae.

queen

Bullet ant workers collecting sugary liquids (honeydew) dropped from planthopper insects in the trees above.

Army Ant (*Eciton burchellii*)

Description: Workers are 0.7 cm (0.3 in) long. Army ants (subfamily Dorylinae) have extremely small eyes. They are most easily identified by their behavior: look for a long procession of ants running very rapidly over the ground. The defining feature of members of the genus *Eciton* is a pair of pointed projections at the back end of the thorax (propodeum). There are two subspecies: *E. burchellii parvispinum* (all black, Pacific slope) and *E. burchelii foreli* (brownish abdomen, Caribbean lowlands). **Natural history**: Army ants are voracious predators that mostly specialize on the larvae and pupae of other ants (these are the white blobs that army ants carry in their mouths), although *Eciton burchellii* captures virtually any small animal that fails to flee from its path. While army ants can sting if you stand in their trail, people living in rural areas usually welcome an army-ant invasion of their house, for when the ants depart after a few hours, the house is free of vermin. Army ants are unusual in at least three respects. First, they always hunt in large groups, which allows them to overwhelm the defenses of other ant colonies. Entering the nest of another ant and plundering the offspring is no easy task, but army ants succeed by rapidly recruiting large numbers of attacking ants. Second, rather than living in a permanent nest, colonies are nomadic; they live in temporary bivouacs (the ants form hanging clusters that contain the queen and her brood) inside hollow trees or under logs. The colony migrates once it has depleted the food in a given area. Third, queens are permanently wingless; thus, when a new queen leaves a colony to form a new colony, she must walk rather than fly. She is joined by half the workers, with the other half choosing to stay with the maternal queen. This unusual manner of colony reproduction ensures that a new colony is never so small that it is incapable of marauding the nests of other ants. The males are also unusual in that they must locate a colony that contains a new queen (in other species of ant, the male searches for virgin queens that have left their maternal colony); when the male finds a new colony, he loses his wings and enters the colony, assuming the workers allow him to do so. When the colony contains hungry larvae, it must move to a new location every night to seek out food to sustain them, during which time the enormous queen is carefully protected by a large entourage of workers. When the larvae pupate, the queen begins laying thousands of eggs and the colony now remains in the same bivouac each night. During the 15-day nomadic phase, the colony makes daily raids but during the 20-day stationary phase raids are conducted on only about half of the days. The massive raids of *E. burchellii* blanket the ground, and nearly everything in the path of the marauding ants must flee or succumb. The escaping insects make easy pickings for several species of birds and parasitic flies that regularly follow these army ants. **Related species**: There are more than 50 species of army ant in Costa Rica, seven of which belong to the genus *Eciton*.

Eciton burchellii foreli capturing a paper wasp (*Polistes major*).

Eciton burchellii parvispinum soldier. Note the huge mandibles.

94

Eciton burchellii parvispinum workers. Each night they form a large bivouac; during the day they continue their rapid march (inset).

Acacia Ant (*Pseudomyrmex flavicornis*)

Description: Workers are 0.7 cm (0.3 in) long. Acacia ants live exclusively on bullhorn acacia shrubs (*Vachellia*). Members of the genus *Pseudomyrmex* have a long, slim body with relatively large eyes. This species is black, whereas many of the other acacia ants are light orange. **Natural history**: The acacias provide the ants with nutritious food packets (Beltian bodies) produced on the tips of young leaflets as well as extrafloral nectar secreted from glands on the leaf petioles. The acacia shrubs also provide the ants with a place to nest (inside the swollen thorns); the ants, in turn, use their mandibles and very painful sting to aggressively defend the acacias against herbivores. They also nip off encroaching vines and make circular clearings on the ground around the base of the acacia, thereby discouraging other ants from visiting the shrub. *Pseudomyrmex flavicornis* is usually found on *Vachellia collinsii*. In his 1874 book, *The Naturalist in Nicaragua*, Thomas Belt mentions this ant (under the name *P. bicolor*). **Related species**: *Pseudomyrmex* is the only genus in the subfamily Pseudomyrmecinae that is present in Costa Rica. There are more than 50 species in the country; only four of these protect bullhorn acacias.

Pseudomyrmex flavicornis on *Vachellia collinsii* (Fabaceae). Note the holes in the thorns.

Pseudomyrmex spinicola at extrafloral nectary on *Vachellia collinsii*.

Cecropia Ants (*Azteca* species)

Description: Workers are 0.4 cm (0.2 in) long. Species of *Azteca* can be identified by their anise-like odor and heart-shaped head (in frontal view). In Costa Rica, five species live exclusively in *Cecropia* trees. **Natural history**: As the name suggests, Cecropia Ants nest inside the naturally hollow trunks of *Cecropia* trees. In addition to providing a place to nest, the plant also provides the ants with nutritious food packets (Mullerian bodies), resembling miniature rice grains, that project from felt pads at the base of each leaf petiole; generally, you will only see Mullerian bodies on young saplings that have not yet been colonized by ants (on older plants, the ants harvest the Mullerian bodies as soon as they are formed). In return, the ants protect the plant from encroaching vines and herbivores; to see this in action, slap the trunk of a *Cecropia* tree and a teeming mass of these ants will quickly cover it. Cecropia ants also maintain mealybugs or scale insects inside the nest, from which they obtain sugary excretions (honeydew). **Related species**: Cecropia Ants belong to the subfamily Dolichoderinae, all of which lack a stinger but defend themselves by biting and by secreting a defensive compound that smells like anise. *Azteca* is the largest genus in this subfamily, with 28 species in Costa Rica.

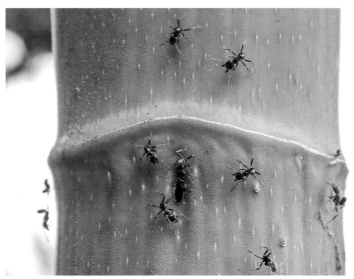

Worker *Azteca* ants in "alarm mode" (with raised abdomen) on *Cecropia* stem; note entrance hole.

Worker ants tending the brood inside a *Cecropia* stem.

Worker harvesting a Mullerian body from the base of the petiole.

Leafcutter Ant (*Atta cephalotes*)

Description: Workers are 0.3–1.4 cm (0.1–0.6 in) long. Leafcutter ants are easily identified by their behavior of cutting and carrying pieces of leaves back to their nest. There are two genera, *Acromyrmex* and *Atta*, both of which have spines on their thorax. *Atta* has huge nests indicated by large mounds of soil. On close inspection, also note its smooth abdomen (as opposed to tuberculate abdomen in *Acromyrmex*). In most of the country, the only species of *Atta* is *A. cephalotes*, but in the southern Pacific region of Costa Rica you will also encounter *A. colombica*. **Natural history**: Leafcutter Ants cut leaves from a variety of plants, sometimes traveling up to 200 m away from the nest to collect them. Their colonies are huge, with the population of a single colony reaching three or four million. The ants use the cut pieces of leaves as a substrate on which to cultivate a fungus (*Leucoagaricus gonglyophorus*) inside their nest. When workers return to the nest, smaller workers chew the leaves into moist globs and add them to the fungal garden, which provides food for the colony. In a mature colony, hundreds of fungal gardens, each in a separate chamber, are connected by a maze of tunnels; underground, the entire nest can encompass an area similar to that of a tennis court. One of their most dangerous enemies is another fungus (*Escovopsis*) that parasitizes the fungal gardens; to combat this parasitic fungus, the leafcutter ants carry on their bodies bacteria (*Pseudonocardia*) that secrete antibiotics. Species of *Atta* are unusual among ants in that the size of the workers varies greatly, depending on how much food the ant receives during its larval stage. Soldier ants, which defend the nest against army ants and other predators, have very large heads; worker ants that cut and carry leaves are medium-sized; another group of worker ants, the miniature workers, are generally confined to the fungal gardens within the nest. The latter sometimes work outside the nest and can be seen riding on a leaf fragment being carried by a medium-sized worker. One of the jobs of these miniature ants is to protect the leaf carrier from attack by parasitic flies (Phoridae) that lay an egg inside the carrier ant's head.

A winged queen

Worker carrying a flower on which rides a hitchhiker ant.

Workers carrying a leaf fragment in their mouth are quite defenseless since the fly lands on the leaf and inserts an egg where the mandible articulates with the head. The miniature "hitchhiker" ant, however, is quite proficient at chasing away flies that attempt to land on the leaf. A new colony requires three or more years to reach maturity, whereupon it produces new queens and males (both of which have wings) once a year, usually a few weeks after the beginning of the rainy season. The males soon die, whereas a mated queen sheds her wings, burrows into the ground to begin a new nest, and starts a garden with fungal fragments that she carries at the back of her mouth. If the queen dies, the colony dies and mortality can be quite high during the first year. If she survives the first few years, the queen has a good chance of living for 20 years or more (probably the longest adult life span of any insect), whereas workers (all unmated females) live for only a month or so. The queen will never again leave the nest and must utilize the sperm she stores inside a special pouch for the rest of her life. **Related species**: Leafcutter ants belong to the largest subfamily of ants, Myrmicinae, which comprises more than 500 of the 900 ant species found in Costa Rica. The two genera of leafcutter ants belong to the subtribe Attina, which includes a dozen genera in total, all of which cultivate fungi. However, ants in the other genera do not cut leaves but instead use fallen flowers, caterpillar excrement, and similar debris as a substrate for their fungal gardens.

Worker ant removing leaf fragment after having made cuts.

Dirt mounds indicate the presence of a leafcutter nest.

Golden Carpenter Ant (*Camponotus sericeiventris*)

Description: Workers are 1.7 cm (0.7 in) long. Carpenter ants (genus *Camponotus*) are moderate to largish ants that lack a stinger and smell like vinegar, due to the formic acid that they secrete from their rear end when distressed. The Golden Carpenter Ant is easily recognized by the short golden hairs covering much of the body. **Natural history**: Carpenter ants derive their name from certain northern temperate-zone species that burrow into wood, not to eat it but rather to build a nest. Costa Rican species vary in their nesting habits, often nesting in rotten wood but never burrowing into sound wood. Some species are common in houses and nest almost anywhere, but Golden Carpenter Ants are generally only found in forests. They are generalist scavengers that harbor endosymbiotic bacteria that probably provide the ants with additional nutrients. **Related species**: Carpenter ants belong to the subfamily Formicinae, which comprises at least 130 species in Costa Rica, about 80 of which belong to the genus *Camponotus*.

Camponotus sericeiventris, probably the most colorful carpenter ant in the country.

Flies
(order Diptera)

As the name *Diptera* suggests, flies have just two wings, the hind pair having been reduced to tiny stubs. In addition to the flies that are a daily presence in our lives, this group also includes mosquitos and many small flies commonly referred to as gnats. The larvae (also called maggots) of most flies lack legs. While the majority of flies are not among the most striking insects that one encounters in the forest, they are the most biologically diverse. Although most larvae are terrestrial, there are more species of aquatic Diptera (midges) than of any other group of insects. Larvae of many species live in leaf litter, some are agricultural pests (fruit flies and some leaf miners), while still others are predators. In terms of medicine, this is the most important group of insects, especially those that transmit deadly microbes such as malaria and dengue; worldwide, mosquitos kill more people than any other group of animals.

Gall Midges (Cecidomyiidae)

These tiny, mosquito-like flies are rarely noticed. However, the galls (abnormal growths) that the larvae form on plants can be readily seen. Other insects and many mites also cause galls but Cecidomyiidae is the predominant group of gall-formers. Each species of gall midge is usually restricted to just a single plant species or to a few related species. Unlike a tumor, a gall stops growing at some point, reaching a characteristic size and shape determined by the species of gall midge responsible for inducing it. Somehow the insect deceives the plant into sending nutrients to the central part of the gall, where the larva resides. Not all gall midges are gall formers. The larvae of many species feed on fungi, a few are predators, and some feed in buds or flowers without inducing a gall. There are an estimated 18,000 species in Costa Rica, though far fewer than 1% have been named. The examples presented here are likely all unnamed species.

Bud gall on *Otopappus verbesinoides* (Asteraceae). It is rosette-shaped (2 cm / 0.8 in wide).

Bud gall on *Acalypha diversifolia* (Euphorbiaceae). Reddish and strawberry-like (1–2 cm / 0.4–0.8 in).

Leaf gall on *Croton guayabo* (Euphorbiaceae). White and hairy (5 mm / 0.2 in).

Leaf gall on *Inga leiocalycina* (Fabaceae). Slender and threadlike (2 cm / 0.8 in tall).

Leaf gall on *Malvaviscus palmanus* (Malvaceae). Ovoid with pointy tip (2 cm / 0.8 in).

Leaf gall on *Eugenia monteverdensis* (Myrtaceae). Resembles grains of rice (7 mm / 0.3 inches long).

Leaf gall on *Psychotria monteverdensis* (Rubiaceae). Resembles sea anemone (3–5 mm / 0.1–0.2 in diameter).

Soldier Fly (*Cyphomyia* species)

map for genus

Description: 1 cm (0.4 in) long. Species in the genus *Cyphomyia* have relatively long, stout antennae and a pair of long spines at the back of the thorax (scutellum). Many have a dark, metallic, blue-black body; females often have a bright yellow head. In males, the eyes occupy nearly the entire head and are usually brown or black; in a few species, however, including the one shown here, the eyes are yellow and thus create the false impression that the male, like the female, has a yellow head. **Natural history**: Males of at least some species hover. Larvae are associated with decaying plant material; they are often found, for example, under the bark of decomposing logs. **Related species**: This is the second largest genus of Stratiomyidae in tropical America; in Costa Rica there are probably at least 20 species in the genus and perhaps 200 species in the family.

female

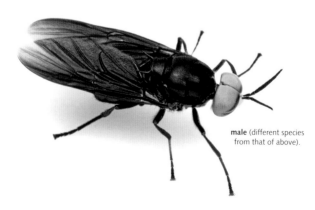

male (different species from that of above).

Robber Fly (*Diogmites* species)

Description: 3 cm (1.2 in) long. Nearly all species of robber fly (Asilidae) have widely separated eyes and a depression on top of the head. Species of *Diogmites* are usually orange to light brown, sometimes with black markings on the thorax or abdomen. **Natural history**: These flies are often seen flying between perching sites on shrubs. Some perch on the ground; *D. littoralis*, for example, perches on the sand on Caribbean beaches. From this perch, they dart out after potential prey, capture it with their legs, stab it with their daggerlike proboscis, and return to their perch. They then inject the prey with saliva, which paralyzes it and liquefies the internal tissues, allowing the fly to suck it dry. Females lay eggs in the soil, which is where the predatory larvae live. **Related species**: There could be more than 20 species of *Diogmites* in Costa Rica; about 6 of these have all-yellow legs, including the individual featured here. They belong to the subfamily Dasypogninae, in the family Asilidae, which in Costa Rica consists of more than 300 species.

Diogmites Robber Fly with prey.

Another genus of Robber Fly, *Mallophora*, above a honeybee, in open sunny field.

107

Long-legged Fly (*Condylostylus* species)

map for genus

Description: 1 cm (0.4 in) long. Long-legged Flies (family Dolichopodidae) have a relatively slender body; most species are shiny green. Species in the genus *Condylostylus* have a depression on the top of the head, between the two eyes. **Natural history**: These flies are commonly seen standing or running along the tops of leaves, fallen tree trunks, or rocks. Adults feed on small, soft-bodied invertebrates in a rather unusual manner: they envelope prey in their spongy proboscis, tearing them open with minute teeth, and then absorbing their bodily fluids. The larvae live in wet leaf litter, mud, or under the bark of rotting tree trunks. They are thought to be mostly predators or scavengers. **Related species**: There are at least 70 species of *Condylostylus* in Costa Rica; the family Dolichopodidae is estimated to have 700 species in the country.

Condylostylus species, dorso-lateral view.

Condylostylus species, dorso-frontal view.

Metallic Green Hoverfly (*Ornidia obesa*)

Description: 1 cm (0.4 in) long. The body is robust and shiny green (even the eyes), with black legs and black spots on the wings. This species might be confused with orchid bees that are green, but it is readily distinguished by its ability to hover in one spot for extended periods of time, then suddenly dart a short distance in any direction. **Natural history**: Members of the family Syrphidae are also called flower flies; indeed, although a few do not visit flowers, *Ornidia obesa* does. The larvae feed in a wide diversity of semi-liquid, decomposing organic material such as rotting fruits and animal dung. Females were observed laying eggs on leaves above a compost pile, and upon hatching the larvae dropped from the leaf. **Related species**: There are 4 species of *Ornidia* in Costa Rica, three of them green, one purple; *O. obesa* is the most common species. There are about 550 species of Syrphidae in the country.

Ornidia obesa, dorso-frontal and ventral views.

Another hover fly, *Allograpta centropogonis*, on *Senecio* (Asteraceae) flowers.

Stilt-legged Fly (*Mesoconius nigrihumeralis*)

Description: 1.6 cm (0.6 in) long. Stilt-legged flies (family Micropezidae) as a group can be distinguished by their extremely long hind legs. The species featured here, like several others in the family, has a white band near the tip of each front leg; by tapping the two front legs they mimic the movement of the white-banded antennae of many ichneumon wasps. The front part of the abdomen is narrow, which also gives it a wasplike appearance. **Natural history**: Species in this genus are rarely found outside of undisturbed cloud forests. Like other members of the family, adults spend a lot of time on the upper surface of leaves, where they feed on honeydew, fallen fruit, and bird droppings. The larvae feed on rotting vegetation, but very little is known about their biology. **Related species**: Most species of *Mesoconius* occur in mountainous regions of the Andes, but 6 species occur in Costa Rica; there are probably at least 100 species of Micropezidae in the country.

Mesoconius nigrihumeralis. Front legs extended forward, resembling the antennae of an ichneumon wasp (p. 79).

Poecilotylus species visiting animal feces.

Tachinid Fly (*Hystricia micans*)

Description: 2 cm (0.8 in) long. Tachinid flies belong to a very diverse family (Tachinidae), ranging in size from smaller than a house fly (0.3 cm / 0.1 in) to much larger (2.5 cm / 1 in). Many of the larger species, such as the one shown here, have long upright bristles on the abdomen. **Natural history**: Adult tachinid flies feed on nectar and other sugary substances, while the larvae are parasites (parasitoids) that live inside the body of another insect. Species of *Hystricia* parasitize caterpillars of wild silk moths (Saturniidae) and tiger moths (Arctiinae). **Related species**: Costa Rica is home to 10 species of *Hystricia* and 2000 species of Tachinidae.

Hystricia micans is usually seen in light gaps and forest edges. Not all tachinid flies are as large and bristly as this species; many resemble a house fly.

Butterflies
and Moths

(order Lepidoptera)

Butterflies and moths are characterized by wings that are covered with scales (the scales give them their colors) and elongated, coiled mouthparts (proboscis) that function like a flexible drinking straw. The immature stages (larvae) are known as caterpillars, which usually have five pairs of prolegs (fleshy stubs) on the abdomen, in addition to three pairs of slender, pointy legs on the thorax. The vast majority of caterpillars feed on plants. Moths generally lack a club at the tip of the antenna, and while most are nocturnal, there are also many diurnal species. Butterflies have clubbed antennae, are nearly all diurnal, and belong to just one superfamily (Papilionoidea). Many butterflies (and several moths) mimic one another, making it easier for predators to remember that a particular color pattern is distasteful, but also complicating matters for people wishing to identify butterfly species.

Moths. This group includes about 90% of all Lepidoptera species, among them a large number of micro-moths that are the size of a mosquito.

Owl Moth (*Automeris postalbida*)

Description: Wingspan 8 cm (3.1 in). The front wings are patterned with reddish-brown and grayish blotches, resembling a dead leaf. The hind wings are normally concealed below the front wings but when threatened the moth extends the front wings to reveal an eyespot on each hind wing. The eyespot is surrounded by a ring, as in the planet Saturn, hence the family name, Saturniidae. **Natural history**: Adults lack functional mouthparts and therefore live for only a week or two. The caterpillars have branched spines that sting if touched; they feed on a wide diversity of plants, usually in a group when young but individually when they are older. **Related species**: There are about 20 species of owl moth (*Automeris*) in Costa Rica; they belong to the subfamily Hemileucinae. Some other genera—*Gamelia* and *Leucanella*, for example—also have an eyespot on each hind wing. The family Saturniidae has about 170 species in the country.

The caterpillar of *Automeris* is covered with stinging spines.

Two views of male *Automeris postalbida*, camouflaged, with wings closed, and startling eye pattern revealed when hind wing is exposed.

Orizaba Silkmoth (*Rothschildia orizaba*)

Description: Wingspan 13 cm (5.1 in). Wings are mostly reddish-brown, but each wing has a translucent triangle and a transverse, wavy white line. A white line between the thorax and abdomen extends as a semicircle onto each front wing. **Natural history**: Like most other silkmoths, the adults have very reduced mouthparts and do not eat. Most large nocturnal moths can detect bat sonar, but silkmoths lack the necessary sensory organs and must rely on other strategies to reduce the risk of being eaten by a bat. Evidence suggests that the hairlike scales on their thorax absorb part of the ultrasounds, thereby reflecting less back to the bat; they also fly erratically. The large green caterpillars feed on the foliage of various trees, including coffee, and upon completing their development they spin a large baglike cocoon in which to pupate. **Related species**: There are 4 species of *Rothschildia* in Costa Rica: this species and *R. erycina*, *R. lebeau*, and *R. triloba*. They belong to the family Saturniidae, subfamily Saturnine.

This male Orizaba Silkmoth was attracted to lights at night and ended up the following day on this door. Inset shows the larva.

Hawk Moth (*Xylophanes crotonis*)

Description: Wingspan 9–10 cm (3.5–3.9 in). In general, hawk moths are easily recognized by their cigar-shaped bodies; long, pointy, narrow wings; and hummingbird-like flight. This species is yellowish- or greenish-brown with longitudinal lines on the front wings and a flame-like cream color pattern on the hind wings. **Natural history**: Adults have excellent night vision and feed on nectar with their extremely long tongues; they are important pollinators of various night-blooming flowers. Species in the genus *Xylophanes* possess tiny structures on each side of the mouth that allow them to detect the ultrasounds produced by bats. The caterpillars, like those of most other hawk moths, have a spine that projects from the rear end (hence the name *hornworms*); they are pale blue with tiny yellow spots and a blue eyespot. The caterpillars feed on certain plants in the coffee family (Rubiaceae). **Related species**: *Xylophanes* is the largest genus of Sphingidae in Costa Rica, with more than 30 species; the family has a total of 150 species in the country.

Xylophanes crotonis hanging from twig. Note how the tip of the abdomen and tips of the wings are pointed, which are characteristics of most hawk moths.

Larva of unidentified *Xylphanes* feeding on *Hamelia patens* (coffee family, Rubiaceae). Note the hornlike projection at the rear end.

Eumorpha vitis, a species in the same family.

Urania Swallowtail Moth (*Urania fulgens*)

Description: Wingspan 10 cm (3.9 in). These black, day-flying moths resemble swallowtail butterflies, but they can be distinguished by their unique iridescent, green-striped color pattern. **Natural history**: Adult Urania moths feed primarily on nectar, mostly from white flowers of legumes and other trees; males are highly attracted to flowers of *Inga*. In certain years, large numbers of these moths migrate throughout the country, flying less than ten meters (33 feet) above the ground and traveling at about 20 kph (12 mph). These migrations apparently occur when caterpillars encounter fewer of their host plants (certain members of the spurge family, Euphorbiaceae). The caterpillars sequester noxious chemicals from their host plants. **Related species**: This is the only member of the subfamily Uraniinae (family Uraniidae) in Costa Rica, although strays of another species, *U. leilus*, from South America, may also occur. There is another subfamily (Epipleminae), with about 40 species in the country, but they are nocturnal and look completely different. The similarly sized and shaped moths *Sematura* are brown with white bands and belong to the family Sematuridae.

Urania Swallowtail Moth on leaf, with head pointed downward.

Black Witch Moth (*Ascalapha odorata*)

Description: Wingspan measures up to 16 cm (6.3 in). In flight, this very large, dark-colored moth could almost be mistaken for a bat. In both sexes, the upper surface of the wings is mottled brown; females have a diagonal white stripe. Near the middle of the front margin of each front wing there is a dark, comma-shaped mark that is iridescent. **Natural history**: During the day these large moths hide in sheltered locations. They fly out at night to search for food (rotting fruits) and also to mate. The caterpillars feed at night on the leaves of various woody legumes; they hide on trunks during the day. **Related species**: This is the only species of *Ascalapha* in Costa Rica; it belongs to the family Erebidae, subfamily Erebinae. *Ascalapha* was previously classified in the family Noctuidae.

Male Black Witch Moth on leaf.

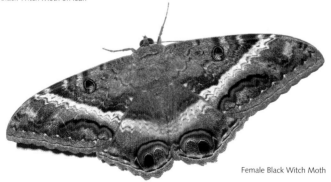

Female Black Witch Moth

119

Druce's Tiger Moth (*Hypocrita drucei*)

Description: Wingspan 6 cm (2.4 in). Front wings dark with a diagonal white band and red spots. Abdomen and hind wings show bluish iridescence with pink and white spots. Abdomen reddish below. **Natural history**: This species is nocturnal but can be found resting on foliage during the day. When disturbed, adults expose the hind wings and pump their abdomen to extrude a foul-smelling yellowish froth from each side of the thorax. The showy coloration probably acts as a warning sign. Caterpillars feed on plants in the genus *Dichapetalum* (Dichapetalaceae). **Related species**: There are at least 6 species of *Hypocrita* in Costa Rica; they belong to the family Erebidae, subfamily Arctiinae (at least 600 species in the country).

Hypocrita drucei attracted to light at night and exuding foul-smelling froth (inset).

Blue Tunic Tiger Moth (*Cyanopepla scintillans*)

Description: Wingspan 4 cm (1.6 in). Body black with extensive blue reflections. Front wing black with a large triangular-shaped red spot and narrow red border at the apex. **Natural history**: These moths fly during the day, fairly low over the ground. Males visit certain plants in the aster family (Asteraceae), borage family (Boraginaceae), and dogbane family (Apocynaceae) in order to obtain defensive compounds (pyrrolizidine alkaloids) that they then pass along in their sperm packet to the female. The caterpillars feed on certain grasses such as *Lasiacis*. They are mostly black with white hairs. **Related species**: There are at least 5 species of *Cyanopepla* in Costa Rica, all with a color combination of black, metallic blue, and red (or orange) patches; they belong to the family Erebidae, subfamily Arctiinae (at least 600 species in the country).

Cyanopepla scintillans feeding on nectar of *Senecio* (Asteraceae).

Butterflies. These insects make up roughly 10% of all Lepidoptera species and just one of more than 30 superfamilies. They can be distinguished from nearly all moths by having clubbed antennae.

Swallowtail Butterflies (Papilionidae)

Swallowtails can be distinguished from other butterflies by their habit of fluttering their wings while feeding. As the name suggests, many have a "tail" on each hind wing. However, not all swallowtails have tails, and note the some species in other families also have tails. Swallowtail caterpillars are unique in having a forked structure (osmeterium) at the front end of the thorax, which they extend when disturbed, thereby dispersing defensive compounds. There are 41 species in Costa Rica.

Great Kite-swallowtail (*Protesilaus protesilaus*)

Description: Wingspan 7.5 cm (3 in). Like other members of the tribe, this species has relatively short, upturned antennae and a semi-translucent appearance. It has a very long, narrow tail on each hind wing. The color is pale yellowish to a very faint bluish, with transverse black bars; the short middle bar on the front margin of the front wing is triangular. **Natural history**: Fast-flying butterflies that prefer bright sunshine. Adults perpetually fly flower to flower for nectar. The caterpillars in this tribe feed primarily on plants in the soursop family (Annonacecae). **Related species**: The only other species of *Protesilaus* in Costa Rica is *P. macrosilaus*. Both belong to the tribe Leptocirini.

Individual visiting flowers of *Verbesina fraseri* (Asteraceae).

Androgeus Swallowtail (*Heraclides androgeus*)

Description: Wingspan 13.5–14 cm (5.3–5.5 in). Overall, males and females are very different. Both have a tail on each hind wing, though it is more pronounced in the male. On male, upper side of wings yellow with black borders but without rows of yellow spots. The female is black with iridescent blue-green on the upper side of the hind wings and with red spots on the undersides. **Natural history**: Both sexes consume nectar from flowers of diverse plants. Males patrol with slow, gliding flight for receptive females; males occasionally seen on riverbanks taking minerals from wet soil or sand and in territorial flight in relatively open areas. Caterpillars resemble bird droppings and feed on the foliage of citrus and other plants in the same family (Rutaceae).

Related species: *Heraclides* is often classified as a subgenus of *Papilio*, but recent molecular evidence suggests that it merits the status of a separate genus. There are 8 species of *Heraclides* in Costa Rica; except for one species, the caterpillars feed on plants in the citrus family (Rutaceae).

Male patrolling his territory.

Male taking minerals from wet soil.

Female visiting *Stachytarpheta frantzii* (Verbenaceae).

Polydamus Swallowtail (*Battus polydamas*)

Description: Wingspan 10–11 cm (3.9–4.3 in). The upper side of the wings are black with a broad yellow band near the outer margin. The undersides are similar, but the hind wings have a row of red spots on the outer margin. **Natural history**: Members of *Battus* and other genera in the group are often known as the pipevine swallowtails because their caterpillars feed on pipevines (Aristolochiaceae), from which they sequester defensive compounds that are passed on to the adults. Adults feed on nectar. **Related species**: There are 6 species of *Battus* in Costa Rica; they belong to the tribe Troidini, which also includes the genus *Parides* and the spectacular birdwings of Southeast Asia.

Individual visiting flowers of *Stachytarpheta frantzii* (Verbenaceae).

Skippers (Hesperiidae)

Skippers are named for their rapid, erratic flight. Other characteristics include a hooked club at the tip of the antenna, broad head, stout body, and relatively small wings. In Costa Rica there are 6 subfamilies of skippers; included here are species from 3 of these.

Dorantes Longtail (*Cecropterus dorantes*)

Description: Wingspan 3.5–4 cm (1.4–1.6 in). Hind wing with a very long and quite broad tail. Almost completely brown, without greenish or bluish reflection; front wing with clear pale spots; underside of hind wing with dark brown spots. **Natural history**: Often associated with disturbed and open habitats. Adults feed on nectar and occasionally from bird droppings. Caterpillars feed on leaves of herbaceous legumes (Fabaceae). **Related species**: Until recently, this species was placed in the genus *Urbanus*. There are 13 species of *Cecropterus* recorded in Costa Rica; they belong to the subfamily Eudaminae.

Cecropterus dorantes perched on vegetation.

Urbanus viterboana, a related species.

Two-barred Flasher (*Telegonus fulgerator* complex)

Description: Wingspan 5–6 cm (2–2.4 in). Most of the body and the base of wings are iridescent blue; rest of wings black with a transverse white stripe in the middle of each front wing. **Natural history**: Adults perch on top and under surface of leaves; they feed on nectar and bird droppings. Males are territorial, especially during the morning and in light gaps. Caterpillars feed on the leaves of trees in the legume (Fabaceae) and other families; they are black with a yellow to orange transverse stripe on each segment. **Related species**: Until recently, this species was placed in the genus *Astraptes*. Based on DNA evidence, the *T. fulgerator* complex is hypothesized to consist of 10 species in Costa Rica that are currently not distinguishable by other means. In addition to this species complex, there are at least 11 other species of *Telegonus* in the country; they belong to the subfamily Eudaminae.

Two-barred Flasher basking on a leaf. Monteverde, 1530 m (5020 ft).

Creon Skipper (*Creonpyge creon*)

Description: Wingspan 5.5 cm (2.2 in). Body black to dark blue; wings iridescent blue with a small bright red spot at the back of each hind wing. **Natural history**: Adults feed on nectar, often from white flowers, and decomposing fruits. Occasionally seen basking (sunbathing) on upper side of leaves. Caterpillars feed on *Dendropanax* (Araliaceae); when not feeding they hide inside a folded-over piece of the leaf; they are black with an orange circle on the side of each segment. **Related species**: This is the only species in the genus, which belongs to the subfamily Pyrrhopyginae.

Individual basking at about 5 m (16 ft) from the ground.

Zabulon Skipper (*Poanes zabulon*)

Description: Wingspan 3 cm (1.2 in). On male, upper side mostly orange with dark brown margins; underside yellowish with dark spots and margins. On female, upper side dark brown with large pale yellow spots near outer margin of front wing; underside similar but with some reddish tinges on hind wing. When perching, wings are usually kept open halfway. **Natural history**: Adults feed on nectar and minerals on ground; males spend much of the day visiting flowers, perched on leaves, defending their territory, and watching out for females—all of which means that they are seen more often they females. The caterpillars feed on grasses (Poaceae). **Related species**: There are at least 3 species of *Poanes* in Costa Rica; they belong to the subfamily Hesperiinae.

Male basking on a leaf and (inset) feeding on sap from *Philodendron brenesii* (Araceae).

Female resting on ground.

128

Sulphur Butterflies (Pieridae: subfamily Coliadinae)

Sulphurs are fast flying butterflies that are common in open areas. While most are yellow, some are extensively white. Sulphurs feed on nectar and males often congregate at puddles to feed. Many species migrate across mountains and along seacoasts. The surface of the upper wing of males reflects ultraviolet light, which is invisible to the human eye but visible to female butterflies, who use the "quality" of the reflected light in choosing a mate. The pale green caterpillars feed predominantly on leaves of trees in the legume family (Fabaceae). There are 25 species recorded from Costa Rica. Because the natural history of sulfur butterflies varies relatively little between species, the authors do not include separate accounts for each of them, but rather some photos of representative species.

White Angled-sulphur (*Anteos clorinde*) female visiting flowers of *Stachytarpheta frantzii* (Verbenaceae).

Yellow Angled-sulphur (*Anteos maerula*) male feeding on nectar.

Apricot Sulphur (*Phoebis argante*) male imbibing nectar from flowers of *Stachytarpheta frantzii* (Verbenaceae).

Tailed Sulphur (*Phoebis virgo*). This species was previously known as *P. rurina* or *P. neocypris*. Both photos show a female individual.

Cloudless Sulphur (*Phoebis marcellina*). Previously *P. sennae*. Male recently emerged from pupa, drying its wings. The larva had fed on *Senna papillosa* (Fabaceae).

Mimosa Yellow (*Pyrisitia nise*) individual basking on a leaf near the ground.

Tailed Orange (*Pyrisitia proterpia*) individuals feeding on nectar of *Muntingia calabura* (Muntingiaceae).

Boisduval's Yellow (*Eurema arbela boisduvaliana*) visiting flower of *Muntingia calabura* (Muntingiaceae).

133

Disjunct Yellow (*Eurema arbela gratiosa*) in flight after visiting *Cuphea carthagenensis* (Lythraceae).

Barred Yellow (*Eurema daira*). Male and female visiting flower of *Tridax procumbens* (Asteraceae); the male opens his wings to court the female.

134

Tropical Yellow (*Eurema xantochlora*) adult recently emerged from the pupa and drying its wings; accompanied by pupae, some of which are empty. The caterpillars fed on *Senna papillosa* (Fabaceae).

Metalmark Butterflies (family Riodinidae)

Metalmarks are small to medium-sized butterflies that vary greatly in wing form and color. The two front legs of males are very small and are not used while perching or walking. They share this characteristic with brush-footed butterflies (Nymphalidae), in which both sexes have small front legs. Many metalmarks perch with their wings extended horizontally at the sides of their body, or with the wings partially open, often while on the undersides of leaves.

Deep-blue Eyed-metalmark (*Mesosemia asa*)

Description: Wingspan 3 cm (1.2 in). On male, upper side iridescent blue with black margins; front wing has more black, including a relatively large black eyespot in the middle of the front margin. Underside brown with a black eyespot on each wing. Female is brown; front wings with diagonal white band and a black eyespot. **Natural history**: Adults are frequently seen near open streams, waterfall areas, and relatively wide roads. While perching on leaves, they occasionally make jerky movements as they maintain wings halfway open. Males and females feed on nectar, especially from tiny white flowers. Caterpillars feed on certain plants in the coffee family (Rubiaceae: *Psychotria*); they are green with scattered hairs and a rounded lobe on each side of each segment. **Related species**: There are 14 species of *Mesosemia* in Costa Rica. They all have the eyespot on their front wings and belong to the tribe Mesosemiini. *M. asa* frequently shares the same habitat with *M. grandis*.

Between brief flights, a male perches along the side of a road.

136

Blue-winged Eurybia (*Eurybia lycisca*)

Description: Wingspan 4 cm (1.6 in). Upper side dark brown with large, iridescent blue area on hind wing; underside light brown; both sides of front wing with a blue eyespot surrounded by a yellow ring. The eyes are deep metallic blue. **Natural history**: Adults usually perch on the underside of leaves. They have a very long proboscis and can therefore reach nectar in deep, tubular flowers; individuals are often seen feeding from flowers of *Calathea* spp. (Marantaceae). Males engage in territorial and courtship behavior at dusk and probably at dawn. The pale yellow-green caterpillars feed inside the flowers of calatheas and secrete substances that attract ants. **Related species**: In Costa Rica there are 5 species of *Eurybia*, most of them possessing the eyespot on their front wings. This is the only genus in the tribe Eurybiini in the country.

Male visiting small tubular flowers of *Psiguria warscewiczii* (Cucurbitaceae).

Stoll's Sarota (*Sarota chrysus*)

Description: Wingspan 2.5 cm (1 in). Hind wing with two long tails. Upper side dark brown in males, more grayish in females, with 4–5 white spots on the front wing. Underside with a complex pattern of mostly reddish-brown and white, with some black lines and pale bluish, silvery-gray areas. **Natural history**: Although eye-catching, these butterflies can be a challenge to see. They are most active in the early to middle afternoon in relatively dark areas of forest understory. Adults feed on nectar. The caterpillars feed on the liverworts and mosses that grow on the upper surface of old leaves; females drag their abdomen over the leaf in order to cover their eggs with detritus. **Related species**: There are at least 10 species of *Sarota* in Costa Rica; they belong to the tribe Helicoptini.

Female *Sarota chrysus* in understory. Note the hairy legs, characteristic of the genus.

Sixola Metalmark (*Calephelis sixola*)

Description: Wingspan 2 cm (0.8 in). Outer margin of front wing sinuous, with apex ending in a curved point. Upper side brown with a transverse dark band in the middle and two thin silvery lines closer to the outer margin. Underside light orangish-brown with several transverse lines. In the dry season, individuals tend to be lighter colored. **Natural history**: Adults fly near ground visiting small flowers along roads, rivers, and other open areas. The caterpillars of this species are unknown. *Calephelis* caterpillars in general are hairy white and feed on plants in the aster family (Asteraceae). This species is endemic to Costa Rica. **Related species**: There are at least 15 species of *Calephelis* recorded from Costa Rica, many of them not easily identified to species level; they belong to the subfamily Riodininae, tribe Riodinini.

Male perched on low vegetation along an open, sunny roadside.

Male with underside of wings visible.

139

Red-bordered Pixie (*Melanis pixe*)

Description: Wingspan 4 cm (1.6 in). Both sides of the wings are black, with a red spot at the base of each wing and a row of red spots on the outer margin; the tip of the front wing is orange. **Natural history**: Adults feed on nectar; sometimes, many individuals visit flowers on a single tree. When at rest, they usually perch on the underside of leaves. Caterpillars feed on the leaves of certain legume trees, including *Inga* (Fabaceae). Despite its common name, this species is a metalmark. **Related species**: There are 3 species of *Melanis* known from Costa Rica. The other two are *M. cephise* and *M. electron*. They belong to the subfamily Riodininae, tribe Riodinini.

Individual visiting flowers of *Acacia angustissima* (Fabaceae).

Blues and Hairstreaks (Lycaenidae)

Most members of this family are quite small, and many have one or more hairlike tails and/or a small, colorful eyespot on each hind wing tip (rear end). When perched with their wings in a vertical (closed) position, they often rub their hind wings together, creating the illusion of a head at the wing tip, which may present a false target for predators. There are two subfamilies in Costa Rica: Polyommatinae ("blues," a few species) and Theclinae ("hairstreaks," the vast majority).

Hanno Blue (*Hemiargus hanno*)

Description: Wingspan 2 cm (0.8 in). Males: upper surface light, iridescent violet with a narrow dark gray line on the outer margins of the wings. Females: upper surface light brown. Both sexes: under surface light gray with numerous small white crescents, 4–5 black spots, and a larger black and orange eyespot on hind wing tips. **Natural history**: Adults fly around in open areas near ground level and feed on nectar. Males are more commonly seen and when perched on vegetation usually open their wings about halfway. The sluglike caterpillars (but with legs) feed on herbaceous legumes (Fabaceae), especially the flower buds. The caterpillars produce secretions that attract ants. **Related species**: The one other species of *Hemiargus* in Costa Rica is *H. ceranus*, which has two eyespots on its hind wing tips. Also note that *H. hanno* occurs in lowland regions of the Caribbean, while *H. ceranus* occurs in the lowlands of both coasts. Both belong to the subfamily Polyommatinae, tribe Polymmatini.

Male visiting a flower of Browne's Blechum, *Ruellia blechum* (Acanthaceae).

Togarna Hairstreak (*Arawacus togarna*)

Description: Wingspan 2.5 cm (1 in). Upper surface of wings white with black margins, with more extensive black on front wings. Lower surface with a zebra pattern, including an orange stripe on the outer margin of both wings and the inner margin of the hind wings. **Natural history**: Generally found relatively low to ground in open, disturbed areas, alternately flying and perching. In flight, the white color of the upper wing stands out. The caterpillars feed on plants in the nightshade family (Solanaceae: *Solanum*); they are stocky, green, and covered with light colored hairs. The genus is named for the Arawaks, the first indigenous peoples encountered by the Spaniards. **Related species**: There are 7 species of *Arawacus* known from Costa Rica; they belong to the subfamily Theclinae, tribe Eumaeini. A similar looking species, *A. linocides*, occurs on the Pacific slope.

Note that the hind wing tail area is black with small white spots, resembling a head with antennae.

Strand's Groundstreak (*Calycopis orcillula*)

Description: Wingspan 2 cm (1 in). Has 2 long, thin tails on each hind wing. Upper surface brown, but female hind wing pale, iridescent blue. Under surface pale grayish brown, with darker transverse line near outer margin and on margin. Rear part of hind wing with orange and black spots surrounded by white lines. **Natural history**: Adults are usually found perched on vegetation, showing the "false-head behavior," which involves rubbing their hind wings together. They fly in relatively dark understory, typically low to ground, and feed on nectar. Caterpillars in this genus and related genera are unusual in that they feed on detritus such as fallen flowers on the forest floor; females lay their eggs on dead leaves and twigs on the ground. **Related species**: There are about 20 species of *Calycopis* known from Costa Rica; they belong to the subfamily Theclinae, tribe Eumaeini. *C. cerata* is another species that is common in Caribbean lowlands.

Male rubbing hind wings together.

White-tipped Cycadian (*Eumaeus godarti*)

Description: Wingspan 5–6 cm (2–2.4 in). Males are usually smaller than females. Upper surface dark bluish-green with iridescence; hind wing outer margin shows pale metallic green. Lower surface black with pale metallic green spots on the outer margin of the hind wing. Tip of abdomen and adjacent area of hind wings orange-red. **Natural history**: Adults fly relatively low to the ground in the understory, where it is usually dark and where their host plants are found. Occasionally seen visiting flowers on trees in open areas. Caterpillars, reddish with transverse pale yellow bands, feed in groups on leaves and cones of native cycads (*Zamia* spp.). **Related species**: This is the only species of *Eumaeus* known from Costa Rica. It belongs to the subfamily Theclinae, tribe Eumaeini.

Male *Eumaeus godarti*

Female is perched on the host plant *Zamia skinneri* (Zamiaceae).

Brush-footed Butterflies (Nymphalidae)

This is the largest family of butterflies in the world. Note how they walk on just the middle and hind legs—the front legs are small, hairy (like a brush), and held against the body. Costa Rica is home to 10 subfamilies, including some popular groups like Morphos, Glasswings, and Passion-vine Butterflies.

Milkweed Butterflies (subfamily Danainae: tribe Danaini)

Milkweed butterflies are primarily an Old World (Indo-Australian) group with only 13 species in the New World, 6 in Costa Rica.

Monarch Butterfly (*Danaus plexippus*)

Description: Wingspan 8–9 cm (3.1 – 3.5 in). Upper and undersides of wings are orange with black veins and margins; also note small white spots on the margins. **Natural history**: In Costa Rica, the Monarch Butterfly does not migrate. Adults feed on nectar and males gather defensive compounds by visiting certain plants, just like their close relatives, the Glasswings (p. 146). The caterpillars have yellow, white, and black rings around the body and a pair of long, soft tentacular structures at each end. They feed primarily on milkweed and other members of the dogbane family (Apocynaceae), from which they sequester defensive chemicals, at least some of which are passed on to the adult stage. **Related species**: There are two other species of *Danaus* in Costa Rica: *D. eresimus* and *D. gilippus*.

Female visiting milk weed flowers, *Asclepias carassavica* (Apocynaceae); this is one of the caterpillar's host plants.

145

Glasswings or Clearwings (subfamily Danainae: tribe Ithomiini)

Most members of this group have elongate wings, a slender abdomen, and a weak antennal club that sometimes has yellow coloration. Many species (not all, despite the name) have transparent to semi-translucent wings that often evoke stained glass. Instead of scales, the wings have tiny well-spaced hairs that allow light to pass right through. To prevent glare, the wing membrane has microscopic "nanopillars" that are so small they can only be seen with a special type of electron microscope. Glasswings typically inhabit the forest understory, usually where it is moist, such as along streams and in ravines. Their flight is slow and fluttery. At the beginning of wet and dry seasons (migratory seasons), many individuals can be seen visiting flowers near the ground. Adults obtain defensive chemicals (pyrrolizidine alkaloids) from certain plants (a few species in the borage, aster, and dogbane families); males pass on extra amounts of alkaloids to females during mating. There are at least 55 species in Costa Rica.

Nero Clearwing (*Godyris nero*)

Description: Wingspan 5.5 cm (2.2 in). Wings are mostly transparent, but with relatively large white patches near the wing tips, and orangish-brown on the wing margins and in the area preceding the large white patch on the front wings. **Natural history**: This species is generally found in forests, at higher elevations. It can be seen flying in open areas during cloudy weather. The caterpillars feed on certain plants in the nightshade family (Solanaceae: *Cestrum*); they rest (and feed) on the upper side of leaves, and construct a silk-lined tubular structure at the leaf tip. **Related species**: There are just 2 species of *Godyris* in Costa Rica; the other species is *G. zavaleta*.

Male visiting flowers of whiteweed, *Agetratum conyzoides* (Asteraceae).

Thick-tipped Greta (*Greta morgane*)

Description: Wingspan 6 cm (2.4 in). Wings are mostly transparent, but front wings have dark brown tips preceded by a long white band. **Natural history**: Very common in disturbed habitats, though abundance at a particular locality sometimes varies due to migration, during which an individual can fly an average of 12 km (7.5 mi) per day. The caterpillars feed on certain plants in the nightshade family (Solanaceae: *Cestrum*). **Related species**: There are 6 species of *Greta* in Costa Rica.

Male imbibing minerals from a bird dropping in forest understory.

Simple Clearwing (*Pteronymia simplex*)

Description: Wingspan 5 cm (2 in). Wings are mostly transparent, with orangish-brown margins and a small white spot. **Natural history**: In open fields, adults visit flowers in the aster family (Asteraceae); they are migratory. Caterpillars are green and feed on certain plants in the nightshade family (Solanaceae: Solanum). Pupae are metallic golden-brown. **Related species**: *Pteronymia* is the largest genus of glasswing in Costa Rica, which has at least 11 species.

Males and females of *Pteronymia simplex* visiting whiteweed flowers, *Ageratum conyzoides* (Asteraceae).

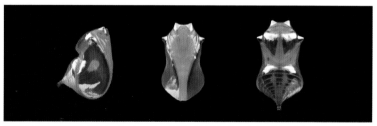

Pupae of related species, *Pteronymia artena*; from left to right, lateral, ventral, and dorsal views.

Polymnia Tigerwing (*Mechanitis polymnia*)

Description: Wingspan 6.5 cm (2.6 in). *Mechanitis* is one of several genera of Ithomiini that does not have transparent wings; species in this genus belong to the tiger-striped mimicry complex, which includes various butterflies (and even a few moths). This species is quite variable but has dark-colored front wings with light yellow spots. **Natural history**: This is the most common Ithomiini in Costa Rica, occurring in both shaded forests and open areas. Caterpillars feed on certain plants in the nightshade family (Solanaceae: *Solanum*). **Related species**: There are 3 species of *Mechanitis* in Costa Rica: this species, *M. lysimnia,* and *M. menapis.*

Male in open field, on whiteweed, *Ageratum* (Asteraceae).

149

Turquoise-banded Shoemaker (*Archaeoprepona amphimachus*)

Description: Wingspan 8 cm (3.1 in). Upper side of wings dark brown with bright iridescent blue transverse band in middle. Underside brown, paler at base, and hind wing with several small brown dots on the outer margin. **Natural history**: These are extremely fast fliers that have rapid wing beats; they feed mostly on rotting fruits and plant sap. Males often rest on foliage or tree trunks, with head down and wings half open; they periodically chase other males and then return to their original perch. Caterpillars feed on plants in the avocado family (Lauraceae). They are light brown with some light green (especially when younger), with two horns on the head, a prominent hump on the thorax, and a pair of eyespots just behind the hump. **Related species**: There are 5 other species of *Archaeoprepona* in Costa Rica; they belong to the tribe Preponini.

Individual feeding on rotten banana; note the pink proboscis.

Perched male adopts a territorial stance.

Individual in flight.

Most members of this subfamily have one or more eyespots on the underside of the hind wing. Owl butterflies (Brassolini) have a maximum of 2 or 3 eyespots on each hind wing, whereas most morphos (Morphini) have at least 4 (if not they have a short tail on the hind wing). Members of this subfamily do not feed on nectar, but rather on fermented fruits, sap flows from wounds in trees, and fungi. There are 5 tribes in Costa Rica: 9 species of Morphini, 27 of Brassolini, and nearly 80 in the other three tribes, which collectively are known as the satyrs; they tend to be smaller and drabber than other members of this subfamily with some exceptions (see page xvi).

Helenor Blue Morpho (*Morpho helenor*)

Description: Wingspan 11 cm (4.3 in). Note combination of large size, bright iridescent blue wings, and white point in the middle of each eyespot on the undersides of the wings. **Natural history**: Males are frequently seen patrolling along forest edges, whereas females tend to stay in the forest. The bouncy zigzag flight is difficult to follow and upon landing the butterfly disappears (with wings held vertically, the dazzling blue suddenly becomes hidden). Adults feed on overripe or rotting fruit. The caterpillars feed on the leaves of certain trees in the legume family (Fabaceae). Like other caterpillars of morphos, they are multicolored and have prominent tufts of weakly urticating hairs. **Related species**: There are 5 other species of *Morpho* in Costa Rica though not all of them are blue (*M. polyphemus* is white and *M. theseus* pearly, brownish-gray). *Morpho deidamia* is similar to *M. helenor*, but the former has the hind wings with more serrated (wavy) outer margins, and it lacks the short black line on front margin of the upper surface of the front wings.

Caterpillars; last stage seen on the left, penultimate stage on the right.

Note brown color and eye spots on underside of wings.

Individual basking on a fallen tree trunk.

Yellow-edged Owl Butterfly (*Caligo atreus*)

Description: One of the largest butterflies, with a wingspan of 12–13 cm (4.7–5.1 in). Upper side dark, with a broad purple transverse band in the middle of the front wing and yellowish pale brown band along the outer margin of the hind wing. Underside of hind wing with yellow band and two eyespots, one larger and more eye-like than the other. **Natural history**: Adults are most active in the early morning and late afternoon, typically chasing after each other in bouncy flight. They feed on rotting fruits and on fermenting plant sap. The eyespots on the hind wings probably serve to startle predators. The caterpillars feed—often in groups—on leaves of *Heliconia* (Heliconiaceae), banana (Musaceae), *Calathea* (Marantaceae), *Guzmania* (Bromeliaceae), and some species of the palm family (Arecaceae). Glands in front of the thorax secrete a substance that can repel army ants. **Related species**: There are 7 species of *Caligo* in Costa Rica. This species and two other species, *C. telamonius* and *C. brasiliensis*, are commonly seen in local butterfly gardens.

Individual sleeping on underside of leaf in head-down position, which makes the wing pattern more closely resemble the face of an owl.

Last (5th) stage caterpillar resting on its host plant (*Heliconia*).

Close up of the face of a larva.

Passion-vine Butterflies (subfamily Heliconinae: tribe Heliconiini)

These butterflies occur primarily in tropical America. Their name derives from the fact that the caterpillars feed on the leaves of passion vines (genus *Passiflora*, also known as passion flowers). Most species have long front wings, large eyes, and long antennae. Species in the genus *Heliconius* fly relatively slowly because they are usually unpalatable to many predators, a trait that is announced by bright warning coloration. These butterflies are famous for their mimicry, sharing similar color patterns with other species in the genus as well as with species in other groups of butterflies. They consume nectar from various plants, but unlike other butterflies, species of *Heliconius* also feed on pollen by dissolving it with salivary enzymes. Some species also feed from moist soil, urine, and bird droppings.

In some species the caterpillars feed on just one or two species of passion vine, while others feed on several. The host plants have evolved various means of protecting themselves from the caterpillars—most have extrafloral nectaries that attract ants that sometimes eat butterfly eggs, and probably all of them produce defensive chemicals (alkaloids and cyanide-producing compounds). Nevertheless, the caterpillars are capable of detoxifying the plant's chemical defenses, and in most species both caterpillars and adults manufacture their own cyanogenic compounds as a defense against predators. However, some species of *Heliconius* (e.g., *H. charithonia*) have lost this ability and have become chemically dependent upon particular species of passion vine, from which caterpillars sequester certain compounds and then pass them on to the adult stage.

The following 8 species are among the most common in this group to be found in Costa Rica. They are presented in cameo form, without complete species accounts, as they all share very similar biology and behavior if not appearance.

Gulf Fritillary (*Agraulis vanillae*) male visiting flowers of *Stachytarpheta frantzii* (Verbenaceae).

Mexican Silverspot (*Dione moneta*) male feeding on minerals in soil.

Juno Silverspot (*Dinoe juno*).

Late stage *Dinoe juno* larvae feeding on *Passiflora ligularis* (Passifloraceae).

157

Julia (*Dryas iulia*) male resting in a shaded area.

Zebra Longwing (*Heliconius charithonia*) male visiting *Lantana* flowers (Verbenaceae) and female basking (inset).

Montane Longwing (*Heliconius clysonymus*) male visiting *Verbesina* (Asteraceae), and female on *Lantana* (Verbenaceae).

Erato Longwing (*Heliconius erato*) male on *Lantana*.   *Heliconius erato* in sleeping position on dry twig.

Hecale Longwing (*Heliconius hecale*) males visiting bird droppings.

Hecale Longwing (*Heliconius hecale*) female

This subfamily is represented in Costa Rica by just a single genus, *Adelpha*. Adults of different species are often very similar to one another even though their host plants vary considerably.

Smooth-banded Sister (*Adelpha cytherea*)

Description: Wingspan 5 cm (2 in). Upper surface dark brown with marbled pattern; front wing with orange band near outer margin and hind wing with white band near outer margin. Underside light orange-brown with a clear white band in the middle and faint white markings. **Natural history**: Adults feed on nectar, rotting fruits, and moisture from fallen tree trunks or from the surface of foliage. They are often seen along forest edges and their swift flight is probably their main defense against predators. Caterpillars feed on certain plants in the coffee family (Rubiaceae: *Sabicea*). Like other species in the genus, the young caterpillar eats away at the tip of a leaf, leaving a projecting mid-vein; it then constructs a chain of frass (feces and dry leaf pieces) along the mid-vein and rests at the end of it, probably to deter predators and/or to camouflage itself. **Related species**: The genus *Adelpha* includes about 38 species in Costa Rica. *A. cytherea* is one of the smallest species of *Adelpha* in Costa Rica and other species tend to have more pointy wing tips.

Male *Adelpha cytherea*

161

Tracta Sister (*Adelpha tracta*)

Description: Wingspan 5.5 cm (2.2 in). Upper side of wings brown to dark brown, with a long orange band across the front and hind wings. The intensity of the orange color fades away toward the hind wing tip, where it terminates in a small, dark orange spot. Underside light brown to brown, with faint white markings. **Natural history**: This is one of the representative *Adelpha* species in the cloud forest habitat. They are usually seen flying along forest edges, especially females trying to lay eggs on the host plant, *Viburnum costaricanum* (Adoxaceae). The early stage caterpillars feed on the tips of leaves, where they often deposit strings of excrement. **Related species**: A smaller species, *A. leuceria*, has a similar wing color pattern; however, the orange band does not become faint on the hind wing.

Male (on left) mating with female.

Red Cracker (*Hamadryas amphinome*)

Description: Wingspan 6 cm (2.4 in). The upper surface of the wings has a calico pattern consisting of slightly iridescent light blue, dark gray, and white. This pattern often blends in well with the bark of the trees on which they perch. Underside wings dark brown with white patches in front wings; most of the hind wings brownish-orange. **Natural history**: Both sexes spend considerable time resting on tree trunks, with their head down and wings spread open, often high above the ground. Like other species in the genus, males are very territorial and make snapping or cracking sounds when they take flight. Caterpillars feed in groups on certain plants in the spurge family (Euphorbiaceae: *Dalechampia*). **Related species**: There is a total of 9 species of *Hamadryas* in Costa Rica; the Red Cracker (*H. amphinome*) has a color pattern similar to that of the Orange Cracker. This genus is the largest in the tribe Ageronini.

Individual attracted to ripe plantain at a farmhouse.

163

Four-spotted Eighty-eight (*Callicore brome*)

Description: Wingspan 4.5 cm (1.8 in). Upper side black; front wing with transverse yellowish-orange band in center, hind wing with iridescent blue band almost reaching base of wing. Underside of front wing similar to upper side but with additional narrow pale yellow band near apex; hind wing black with four light blue dots in center, surrounded by pale yellow semicircular bands. **Natural history**: Both sexes feed on rotting fruits, and males sometimes visit mud puddles. Caterpillars feed on certain plants in the soapberry family (Sapindaceae). **Related species**: There is a total of 7 species of *Callicore* in Costa Rica; they belong to the tribe Callicorini.

Male feeding on minerals on sandy beach.

Cramer's Eighty-eight (*Diaethria clymena*)

Description: Wingspan 3.5 cm (1.4 in). Upper side black, with a diagonal metallic blue to pale green band on each front wing. The underside of the front wing is red at the base, with transverse black and cream-colored stripes at the apex. The underside of the hind wing is light yellowish-gray with a black figure 88 or 98 surrounded by black circular lines. **Natural history**: This species is often found around (and sometimes in) human habitations and on the walls of buildings. When it perches on a vertical surface, it does so with its head pointing downward. Adults feed on rotting fruits and males are often attracted to wet soil or moist laundry. Caterpillars feed on *Trema* trees (Cannabaceae), a plant that resembles hackberry. **Related species**: There are 4 other species of *Diaethria* in Costa Rica; they belong to the tribe Callicorini.

Male feeding on minerals from floor tiles.

Male basking, showing upper side of wings.

Orange-striped Eighty-eight (*Diaethria pandama*)

Description: Wingspan 4 cm (1.6 in). Upper side brown to dark brown, front wing with transverse yellowish-orange band in center and white dot at apex. Underside of front wing reddish near base, surrounded by black bands; hind wing light gray with faint figure 88 in center surrounded by very fine, curvy black lines. **Natural history**: Males are attracted to damp soil, urine-soaked sand, wet and dry concrete rocks, and dung. They are easily disturbed, perching for just a few seconds, flying off erratically and close to the ground, but returning repeatedly to the same spot. Caterpillars feed on certain plants in the soapberry family (Sapindaceae: *Serjania*). **Related species**: Among the 5 species of *Diaethria* in Costa Rica, this is the only species that lacks metallic bands on upper side of the wings.

The underside of the front wing shows red (upper left). Note white dot on the upper side of the front wing (upper right). Also note figure 88 on underside of hind wing (directly above).

Daggerwings (subfamily Cyrestinae)

These butterflies take their name from the long tail on each hind wing. There is only one genus in the Americas.

Purple-stained Daggerwing (*Marpesia marcella*)

Description: Wingspan 5.5 cm (2.2 in). Males: upper side of wings brown to dark brown with broad orange band in middle leading to hind wing; hind wings with broad purple iridescent band near inner margin towards the tails. Underside pale orangish-brown with narrow white lines. Females: upper side brown, front wings with broad white band in middle. Underside pale brown with faint white lines. **Natural history**: Adults feed on nectar; groups of males often imbibe mineral-laden water in wet sand or mud. While imbibing they open and close their wings. Females spend a lot of time in the canopy, but on overcast days they sometimes come down closer to the ground. Caterpillars feed on certain plants in the fig family (Moraceae). They are brightly colored and have a few long spines along the center of the back and a couple of long projections emanating from the head. **Related species**: There are 8 species of *Marpesia* in Costa Rica.

On male, note upper side of wings with broad orange and purple bands; underside of wings are pale brown with white stripes (inset).

167

Beauties, Checkerspots, and Their Kin (subfamily Nymphalinae)

This diverse subfamily consists of 5 tribes in Costa Rica. Representatives of 3 of these tribes are described below.

Dirce Beauty (*Colobura dirce*)

Description: Wingspan 6 cm (2.4 in). Upper side with dark coloration, except for a transverse yellow band in the middle of the front wing. Underside with a zebra pattern of fine black lines on a white background. **Natural history**: Adults feed on rotting fruits, carrion, and dung. They often perch on tree trunks, head down, a couple of meters or more above the ground. Caterpillars feed on *Cecropia* and defend themselves against the ants that occupy these trees by secreting a repellent and, when they are young, by resting on top of their own excrement. Mature caterpillars are black with yellow branched spines. **Related species**: There are only 2 species of *Colobura* in Costa Rica, the other one being *C. annulata*. This genus belongs to the tribe Nymphalini.

Individual (underside showing) on plant stem with head pointed downward. Last stage larva (inset) of *Colubra* species feeding on *Cecropia obtusifolia* (Urticaceae).

Blomfild's Beauty (*Smyrna blomfildia*)

Description: Wingspan 6–7 cm (2.4–2.8 in). Upper side of wings is mostly reddish-orange in males and brown in females; apex black with three white spots. Underside of wings with a complex pattern of circular creamy white lines and with four eyespots near the outer margin. **Natural history**: Adults are fast flyers and usually spend time in sub-canopy areas of all types of forest. They frequently come down to ground level to feed on rotting fruits and dung. Caterpillars feed on plants in the nettle family (Urticaceae: *Urera*) and are mostly black with branched spines all along the body. **Related species**: The genus *Smyrna* has just 2 species, this and *S. karwinskii*. The later has only been reported from Mexico to Nicaragua. The genus belongs to the tribe Nymphalini.

Male with wings partially open; note reddish-orange upper side.

169

Banded Peacock (*Anartia fatima*)

Description: Wingspan 5 cm (2 in). Outer margin of hind wing wavy. Upper side brown to dark brown, darker at apex, with transverse white to yellowish band, and red and white (yellowish) spots. Underside similar but paler. **Natural history**: Very common in disturbed habitats. Adults feed on nectar. Males defend territories against other males, alternately flying in a slow zigzag low to the ground and perching on leaves. Caterpillars feed on plants in the acanthus family (Acanthaceae); mature caterpillars are black with finely branched spines. **Related species**: There are just 2 species of *Anartia* in Costa Rica, this and *A. jatrophae*, also common in similar habitats. The genus belongs to the tribe Victorini.

Female basking on a leaf.

Rusty-tipped Page (*Siproeta epaphus*)

Description: Wingspan 8 cm (3.1 in). Surface of upper wing dark brown, with apex of front wings rusty-orange and with a narrow white transverse stripe separating these two colors. Lower surface brown with a transverse white stripe. **Natural history**: This species is common in disturbed habitats and forest edges. Both sexes feed on nectar and rotting fruit, while males also seek out minerals from dung and wet soil. The caterpillars feed on the foliage of certain plants in the acanthus family (Acanthaceae). Mature caterpillars are velvety maroon with 3 pairs of yellowish spines on each segment. **Related species**: There are 3 species of *Siproeta* in Costa Rica; *S. stelenes* is green and *S. superba* lacks the rusty-orange on the wing tips and has a broader white band. The genus belongs to the tribe Victorini.

Siproeta epaphus

171

Crimson Patch (*Chlosyne janais*)

Description: Wingspan 5 cm (2 in). The upper and undersides of the front wings are black with small white spots. The upper side of the hind wing is black with a large reddish-orange patch at the base; the lower side of the hind wing has orange and yellow markings. **Natural history**: Common in open, disturbed areas. Adults open and close their wings while feeding on nectar. Caterpillars feed in groups on the undersides of the leaves of plants in the acanthus family (Acanthaceae). They are light gray with black rings, with a circle of spines arising from each black ring. **Related species**: There are 11 species of *Chlosyne* in Costa Rica; they belong to the tribe Melitaeini.

Photos showing underside (top) and upper side (bottom).

Orange-patched Crescent (*Anthanassa drusilla*)

Description: Wingspan 3.5 cm (1.4 in). Upper surface dark brown with pale yellow markings and orange markings at the base of the hind wing; underside similar but paler. Species in this genus have a slight concavity on the outer margin of the front wing and a fine, wavy line near the outer margin of the hind wing. **Natural history**: Common in open areas. Both sexes feed on nectar, while males also seek out wet sand and urine. Caterpillars feed on plants in the acanth family (Acanthaceae). **Related species**: There are 9 species of *Anthanassa* in Costa Rica; they belong to the tribe Melitaeini.

Male on roadside visiting flowers of false buttonweed, *Spermacoce* sp. (Rubiaceae).

173

Black-bordered Crescent (*Tegosa anieta*)

Description: Wingspan 3 cm (1.2 in). Upper surface orange with dark brown borders and a transverse black band near the apex of the front wing that encloses an orange patch. Underside pale orangish-brown with irregular brownish markings. **Natural history**: Adults feed on nectar and occasionally from soil (in gatherings). With their quick and smooth flight, they often chase away other larger butterflies in the vicinity of flowering plants. Caterpillars feed in groups (at least when young) on certain plants in the aster family (Asteraceae). Mature caterpillars are dark with a circle of spines on each segment. **Related species**: There are 3 species of *Tegosa* in Costa Rica. The other species are *T. claudina* and *T. nigrella*; the former species looks very similar to *T. anieta*. The genus belongs to the tribe Melitaeini.

Male resting on ground.

174

Other
Arthropods

The animal kingdom is divided into more than 30 major groups (known as phyla), and the largest of these is Arthropoda, whose members are characterized by an external skeleton that must be periodically molted. The name *arthropod* means jointed leg, another characteristic. In addition to insects, the arthropods include arachnids, centipedes, millipedes, and crustaceans.

Arachnids (subphylum Chelicerata)

The arachnids include scorpions, pseudoscorpions, harvestmen, whip spiders, spiders, mites (by far the most diverse group), and several lesser known groups. Unlike other arthropods, they lack antennae. Except for harvestmen and many mites, arachnids are predators that feed by sucking up the liquefied contents of their prey.

Harvestmen (*Prionostemma* species)

Description: 0.5 cm (0.2 in). The extremely long legs give rise to their other name, *daddy longlegs*. Cellar spiders (Pholcidae) also have very long legs, but the bodies of spiders are divided into two parts and they live in webs; the body of harvestmen is not divided into two parts and they do not live in webs. **Natural history**: During the day, harvestmen in this genus often aggregate in protected sites on vegetation; they can be found, for example, on the trunks of spiny palms or the buttresses of trees. If disturbed, they all bob up and down. At night many of them descend to the ground to search for food—live or dead invertebrates, excrement, fallen fruit, and fungi. Aggregations generally occur in the same sites over long periods of time, sometimes for many years, which is probably a result of the harvestmen leaving chemical markers at their communal roosting sites. **Related species**: There are perhaps a dozen or more species of *Prionostemma* in Costa Rica. They belong to the family Sclerosomatidae (order Opiliones). Species in most other genera have much shorter legs and do not climb trees.

Prionostemma species in aggregation.

Edward's Bark Scorpion (*Centruroides edwardsii*)

Description: Up to 11 cm (4.3 in) in length. Color varies, but this scorpion is usually yellowish-brown with the pincers and tip of the tail darker; also note lines formed by tiny, elevated bumps on the pincers and tail. **Natural history**: Though it also occurs in forests, this is the scorpion that has most successfully adapted to urban areas. It hides in cavities during the day and comes out at night to feed on insects. In courtship, the male and female grab each other by the pincers and perform a prolonged dance, after which the male deposits a sperm packet on the substrate so that the female can retrieve it. Scorpions are unusual among arachnids in that females give live birth to miniature scorpions. The young crawl onto their mother's back, and she carries them around until they have undergone at least one molt. None of the scorpions in Costa Rica has a lethal venom, although it is possible that a given person might have a serious allergic reaction to the sting. **Related species**: Costa Rica is home to 15 species of scorpion, divided into four families. The largest family is Buthidae, of which Edward's Bark Scorpion is a member.

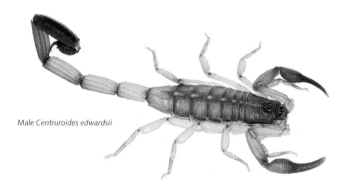

Male *Centruroides edwardsii*

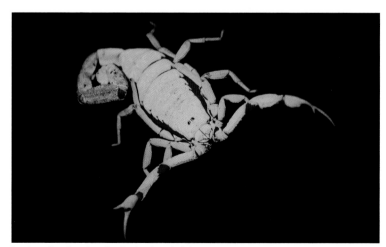

Scorpions often glow when placed under ultraviolet light.

Costa Rican Redleg Tarantula (*Megaphobema mesomelas*)

Description: Up to 6 cm (2.4 in) in length. A large, hairy spider. Black with broad reddish-brown bands on the legs. **Natural history**: Like other tarantulas, this species spends most of its time in burrows waiting for prey to pass in front of the entrance; the hunter is alerted to the presence of prey by the vibrations of silken threads emanating from the opening of the burrow. It is most active at night. Female tarantulas can live for many years and are unusual in that they molt during the adult stage. If threatened, tarantulas are capable of inflicting a painful bite, although their venom is generally not dangerous to humans; they do have, however, irritating hairs on their abdomen. **Related species**: In Costa Rica, there are 3 species in this genus and, overall, more than 40 species of tarantula (Theraphosidae).

Female *Megaphobema mesomelas* at her burrow entrance. Note the thin, white threads of silk.

Tengella Spider (*Tengella radiata*)

Description: 2 cm (0.8 in) in length. Males are slightly smaller than females. Females have very long legs; their front legs can measure 3.5 cm (1.4 in). Light brown with pale yellow lines on the cephalothorax (the body part in front of the abdomen) and pale yellow dots on the abdomen. **Natural history**: This species is quite common and occurs in a variety of habitats, including primary forests, coffee plantations, and roadside embankments. It builds a horizontal sheet web that is turned upward at the edges and narrows into a tubular retreat at the back. The spider hides in the tube waiting for prey; when an insect falls onto the web, it rushes out, bites the prey, and carries it back to the retreat to feed. **Related species**: The genus *Tengella* is known only from Central America and Mexico; *T. radiata* is the only member of the genus (and of the family Zoropsidae) currently known to occur in Costa Rica. Although this species builds webs, it closest relatives (wolf spiders and related spiders) do not.

Tengella radiata waiting in "ready position" near its tubular retreat.

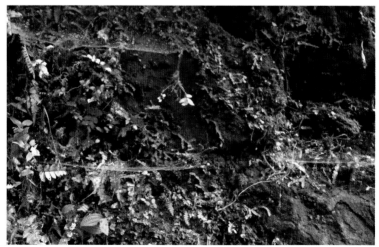

Distinctive webs of *Tengella radiata* along a roadside embankment.

Spinybacked Orbweaver (*Gasteracantha cancriformis*)

Description: Females up to 0.9 cm (0.4 in) in length, 1.3 cm (0.5 in.) in width; males much smaller, only 0.3 cm (0.1 in) in length. The abdomen of females is hard and has 6 peripheral spines. The color is quite variable, but most individuals are whitish, yellow, orange, or red, with black spots. **Natural history**: The immature stages and adult females build orbicular (wheel-shaped) webs that they use for capturing insects. Each night they generally dismantle the web and build a new one, thus ensuring that the threads remain sticky. Mature males dangle from single threads nearby and mating occurs in the female's web. The eggs, about 100–200 of them, are deposited in a silken sac on the underside of a leaf adjacent to the female's web; this sac is 2.0–2.5 cm (about 1 in) long and covered by coarse yellow threads with a longitudinal line of green silk running down the middle. **Related species**: This is the only species in the genus in continental America. Orbweavers in the genus *Micrathena* (about 20 species in Costa Rica) also have a spiny abdomen, but in this genus the body is longer than wide. Both genera belong to the main family of orbweavers, the Araneidae.

Gasteracantha cancriformis

Another genus of orbweaver, *Micrathena* species (ventral view). Note the two large thorny projections.

Golden Silk Orbweaver (*Trichonephila clavipes*, previously *Nephila clavipes*)

Description: Females are among the largest web-building spiders, with a body that measures 2.5–4 cm (1–1.6 in); males are tiny (0.6 cm / 0.2 in). The front of the body is silvery, while the elongate abdomen is tan with yellow spots; the legs are banded with prominent tufts of black hairs (except for the third pair of legs, which lack the bands of black hairs). The tiny males are most easily identified by their mere presence in the web of a female. **Natural history**: The web of an adult female, which can reach a meter or more in diameter, is constructed of golden-colored silk, although the female can adjust pigment intensity in the silk relative to the background colors. The hub of the circular web is not located at the center but toward the top of the web; from there, the female monitors the silken threads for vibrations caused by a struggling insect. While some spiders subdue prey by wrapping it in silk, this species first bites the prey. This means that it cannot deal effectively with large or aggressive prey. After biting, the spider pulls the prey out of the web and carries it back to the hub to feed; larger prey items are often wrapped in silk before being transported to the hub. **Related species**: Previously placed in Nephilidae, now Araneidae.

Female with a male on its underside; note second male lurking in the background. Inset shows dorsal view of female.

181

Orchard Orbweaver (*Leucauge mariana*)

Description: Females 0.5–1.2 cm (0.2–0.5 in) in length, males slightly smaller. The elliptical abdomen has colorful green, yellow, and silvery stripes. **Natural history**: During the rainy season, this species is abundant in the Central Valley and in disturbed habitats elsewhere in the country. Immature stages and adult females build orbicular (wheel-shaped) webs that are usually constructed in a roughly horizontal position. The spider waits for prey on the underside of the hub, using its legs to sense the vibrations produced by an insect striking the web. Males mate with females in the latter's web. Females construct silken sacs on the ground and cover them with debris. **Related species**: There are at least 5 species of *Leucauge* in Costa Rica. They belong to the family Tetragnathidae.

Note the colorful stripes on the abdomen; inset shows ventral view.

Centipedes and Millipedes (subphylum Myriapoda)

This is the smallest group of arthropods. The number on legs on centipedes and millipedes varies from species to species. Centipedes have one pair of legs per segment and are quick predators, whereas millipedes have two pairs of legs per segment and are slow moving consumers of decomposing plant material.

Python Millipede (*Nyssodesmus python*)

Description: Up to 10 cm (4 in) long. Has flat segments and a hard exoskeleton (containing much calcium). Also known as the Large Forest-floor Millipede, it is pale in the middle, bordered by black on each side, with yellow on the side of each segment. **Natural history**: Like many other millipedes, this species feeds on rotting wood and plays an important role in nutrient recycling. When disturbed, it rolls up into a spiral and secretes cyanide. After mating, the male spends several days riding on top of the female to prevent other males from mating with her. Eggs are laid in a cavity in the soil; the newly hatched millipedes have only seven segments but add more segments each time they molt. **Related species**: There are probably about 10 species in this genus in Costa Rica. They are members of the Flat-backed Millipede family (Platyrhacidae).

Nyssodesmus python moving slowly on moist forest floor and (inset) rolled up.

Crabs (subphylum Crustacea)

While several crustaceans occur in freshwater and a few are terrestrial, most live in the ocean. Marine crustaceans vary from barnacles attached to intertidal rocks to microscopic species inhabiting the open sea. Among the few crabs that live on land, the majority must return to the ocean to lay eggs. The latter are usually incubated in the female's body until they are nearly ready to hatch, at which point she releases them into the ocean, usually at night and at high tide. The eggs produce planktonic larvae that look nothing like a crab; after a few weeks to a couple months these immature forms come ashore and transform into juvenile crabs.

Pacific Hermit Crab (*Coenobita compressus*)

Description: Body up to 1.3 cm (0.5 in) in length. Color variable but usually tan. The most obvious characteristic of hermit crabs is that they walk about on just four legs, with most of their body hidden inside an empty snail shell. **Natural history**: The snail shell provides the crab with protection from predators and desiccation. However, it also adds extra weight to carry around and the crab must seek out larger shells to accommodate its growing body. Shells of the right size are often hard to find and so these crabs sometimes resort to killing a snail or stealing a shell from another hermit crab. The name *hermit* is a bit of a misnomer since hermit crabs often forage in groups, feeding on excrement, carrion, fallen fruits, and other detritus. In order to mate, hermit crabs emerge part way from their shell. **Related species**: There is only one other terrestrial hermit crab (family Coenobitidae) in Costa Rica, the Caribbean Hermit Crab (*C. clypeatus*), which occurs on that coast.

Coenobita compressus

Fiddler Crab (*Minuca mordax*)

Description: As on ghost crabs (p. 186), the eyes of the Fiddler Crab are situated at the tips of long stalks. In males, one claw (pincer) is much larger than the other, sometimes weighing half his body weight; as he eats, the rapid movement of the small claw from ground to mouth gives the impression that he is playing the large claw like a fiddle (hence the name). Within a given species, about half of the males have the right claw enlarged while the rest have the left claw enlarged. **Natural history**: Fiddler crabs live in colonies, either in the intertidal zone (the area between low and high tide marks) or the area just above the high tide line. Each crab digs a burrow and defends the area surrounding it. During high tide, they seal their burrows from inside, but as the tide recedes they come out to feed on organic detritus deposited on the sand or mud flats. With their small claw (both claws in the case of females), they scoop sediment into their mouth and then flood the oral cavity with water from the gill chamber; lighter edible particles float to the top and are washed into the gut, whereas the indigestible material sinks to the bottom and is spit out as small pellets. The male waves his large claw to attract females and to threaten or fight other males. **Related species**: In Costa Rica, this is the only species of fiddler crab found on the Caribbean coast, but there are nearly 30 species (in four genera) on the Pacific coast. Fiddler crabs belong to the same family (Ocypodidae) as do ghost crabs.

Male *Minuca mordax*

185

Painted Ghost Crab (*Ocypode gaudichaudi*)
Gulf Ghost Crab (*Hoplocypode occidentalis*)

for both species

Description: 1.5–3 cm (0.6–1.2 in) in diameter. Body (carapace) is square-shaped. The eyes are situated at the tips of long stalks. The Painted Ghost Crab is reddish-orange with numerous whitish dots; each eye has an upward projecting "horn." The Gulf Ghost Crab is dark gray with extensive, irregular whitish markings, and lacks the "horn" on the eye. Juveniles of these two species can be difficult to distinguish. **Natural history**: Ghost crabs get their name from their nocturnal habits (though the painted ghost crab is mostly diurnal) and ability to rapidly disappear down their burrows; *Ocypode* means, appropriately enough, "swift footed." Their widely separated burrows are found from just below the high tide line to as far as 400 m (0.25 miles) inland, with those of juveniles occurring closer to the water line. Painted Ghost Crabs prefer rockier beaches, whereas Gulf Ghost Crabs prefer sandy beaches. Both species feed on dead plant and animal material, as well as small invertebrates buried in the intertidal sand such as mole crabs and small clams. The Gulf Ghost Crab has been reported preying on the eggs and hatchlings of sea turtles. **Related species**: There are just three species of ghost crab in Costa Rica, two on the Pacific (both featured here) and the Atlantic Ghost Crab (*O. quadrata*).

Ocypode gaudichaudi at its burrow. It can retreat into it in an instant.

Hoplocypode occidentalis is well camouflaged against the sand.

Halloween Crab (*Gecarcinus quadratus*)

Description: Body 5–7 cm (2–2.8 in) in length. This crab has a black body with a couple of light-colored spots behind the eyes, purple claws, and bright reddish-orange legs. This color combination is the reason for its two common names, Halloween Crab and Red Land Crab. **Natural history**: The burrows of Halloween Crabs are sometimes quite abundant in Pacific coastal forests and nearby foothills. They feed on leaf litter and seedlings, and by bringing these materials into their burrows the crabs play important roles in both nutrient cycling and seedling survival. While Halloween Crabs usually stay on the ground near their burrows, they occasionally climb trees. They are most active at night and during the rainy season. **Related species**: The only other species of Halloween Crab in Costa Rica, *G. lateralis*, is found on the Caribbean coast. Together with the blue land crabs (*Cardisoma*), they comprise the family Gecarcinidae.

Gecarcinus quadratus

Suggested Reading

DeVries, P.J. 1987. *The Butterflies of Costa Rica and their Natural History, Vol. I: Papilionidae, Pieridae, Nymphalidae*. Princeton: Princeton University Press. 327 pp.

DeVries, P.J. 1997. *The Butterflies of Costa Rica and their Natural History, Vol. II: Riodinidae*. Princeton: Princeton University Press. 288 pp.

Hanson, P.E., and K. Nishida. 2016. *Insects and Other Arthropods of Tropical America*. Ithaca, New York: Cornell University Press. 384 pp.

Hogue, C.L. 1993. *Latin American Insects and Entomology*. Berkeley: University of California Press. 594 pp.

Hölldobler, B., and E.O. Wilson. 2009. *The Superorganism: The Beauty, Elegance, and Strangeness of Insect Societies*. New York: W.W. Norton & Company. 522 pp.

Levi, H.W., and L.R. Levi. 2001. *Spiders and their Kin: A Fully Illustrated, Authoritative and Easy-to-Use Guide*. New York: Golden Guides from St. Martin's Press. 160 pp.

Naskrecki, P. 2017. *Hidden Kingdom: The Insect Life of Costa Rica*. Ithaca, New York: Cornell University Press. 207 pp.

Paulson, D., and W. Haber. 2021. *Dragonflies and Damselflies of Costa Rica: A Field Guide*. Ithaca, New York: Cornell University Press. 401 pp.

Sverdrup-Thygeson, A. 2020. *Extraordinary Insects: Weird. Wonderful. Indispensable. The ones who run our world*. London: Mudlark, Harper Collins. 320 pp.

Photo Credits

Except for the following, all photographs are by Kenji Nishida:

p. 5 (illustration): Cope Arte
pp. 15, 36 (*Apiomerus vexillarius*): Piotr Naskrecki
p. 88: Aiko Kimura
pp. 18 (Championica frontal view) and 185: Ángel Solís

Captions for full-page photographs:

p. ii: *Lissomus* sp. (Elateridae)
p. vi: Arctiinae (Erebidae)
p. viii: gold form of *Chrysina chrysargyrea* (Scarabaeidae)
p. x: *Anapolisia maculosa* (Tettigoniidae)
p. 6:   *Pierella helvina* (Nymphalidae: Satyrinae)
p. 190:   *Hoplopyga liturata* (Scarabaeidae)
p. 192:   Mantispidae (Neuroptera)

Index

Abana gigas 48
Acacia Ant 96
Acanthaceae 141, 170-173
Acrididae 21-22
Acrocinus longimanus 63
Adelpha cytherea 161
Adelpha tracta 162
Adoxaceae 162
Aeshnidae 11
Agaonidae 82
Agelaia panamensis 80
Agraulis vanillae 156
Agrilus 57
Allograpta centropogonis 109
Alurnus ornatus 68
Ampelophilus truncatus 21
Ampulex 87
Ampulicidae 87
Anapolisia maculosa x
Anartia fatima 170
Anchemoia Sphinx Moth 1
Androgeus Swallowtail 123
Anisolabididae 15
Anisoscelis alipes 40
Annonaceae 122
Anteos clorinde 129
Anteos maerula 129
Anthanassa drusilla 173
Ants 49, 93-100, 137, 141, 156
Apidae 90-92
Apiomerus vexillarius 36
Apocynaceae 1121, 145-146
Apricot Sulphur 130
Aquatic insects 8-14, 32-33
Araceae (aroids) 128
Arachnids 176-182
Araliaceae 24, 127
Araneidae 180-181
Arawacus togarna 142
Arctiinae vi
Archaeoprepona amphimachus 150-151
Arctiinae 111, 120-121
Argia anceps 9
Argia cupraurea 9
Arilus gallus 35
Aristolichiaceae 124
Army Ants 94-95
Ascalapha odorata 119
Asilidae 107
Aspisoma 59-60
Assassin bugs 35-37
Asteraceae 19, 21, 76, 102, 109, 121-122,
    134, 139, 146, 148-149, 159, 174

Astraptes 126
Atta cephalotes 98-99
Auchenorrhyncha 42-50
Automeris postalbida 113-114
Avocado family 150
Azteca 97

Bacteria 98, 90
Banana 30, 154
Banded Peacock 170
Baridinae 74-75
Bark Mantis 27
Barred Yellow 134
Battus polydamas 134
Bearded Palm Weevil 72
Beauties 168-169
Bee Assassin 36
Bees 83, 90-92, 107
Beetles 52-76
Belt, Thomas 96
Bess Beetle 53
Biblidinae 163-166
Bird Grasshopper 22
Blaberidae 30
Black and White Weevil 75
Black pepper family 75
Black Witch Moth 119
Black-bellied Leptinotarsa 67
Black-bordered Crescent 174
Blattodea 30-31
Blomfild's Beauty 169
Blue Tunic Tiger Moth 121
Blues 141
Blue-winged Eurybia 137
Boisduval's Yellow 133
Boraginaceae 70, 121, 146
Braconidae 78
Broad Wing Katydid 19
Broad-bodied leaf beetles 66-67
Bromeliaceae 154
Brush-footed butterflies 145-174
Bullet Ant 93
Bullhorn Stink Bug 38
Buprestidae 57
Burseraceae 42
Butterflies 123-174

Caddisflies 33
Caenia kirschi 61
Calephelis sixola 139
Caligo atreus 154-155
Callicore brome 164
Calligrapha fulvipes 66

Calopterygidae 8
Calycopis orcillula 143
Camponotus sericeiventri 100
Camouflage 24-25, 27-29, 42, 114, 187
Cantharidae 62
Carcinophora americana 15
Carpenter ants 100
Cassidinae 68-71
Caterpillars (photos of) 1, 3, 113, 115, 117, 152, 155, 157, 168
Caterpillars 35, 81, 99, 111, 113, 115-129, 135-150, 152, 154-157, 161-174
Cecidomyiidae 102-105
Cecropia 97, 168
Cecropia Ants 97
Cecropterus dorantes 125
Ceiba (plant) 57
Centipedes 183
Centruroides edwardsii 177
Cerambycidae 64-65
Cercopidae 45
Chagas disease 37
Champion's Katydid 18
Championica montana 18
Charaxinae 150-151
Charidotella egregia 69
Chauliognathus heros 62
Checkerspots 172-174
Chelicerata 176-182
Chemical defenses of insects 39, 52, 60-62, 66, 97, 118, 121-122, 124, 145-146, 156
Chlorida cincta 65
Chlosyne janais 172
Choeradodinae 28
Choeradodis rhombicollis 28
Cholus costaricensis 76
Cholus Weevil 76
Chrysina chrysargyrea viii, 56
Chrysobothris delectabilis 57
Chrysomelidae 66-71
Chrysomelinae 66-67
Cicadas 44
Cicadellidae 46-48
Cicadidae 44
Cicindelidae 52
Citrus family 43, 123
Clearwing butterflies 146-149
Click beetles 58-59
Cloudless Sulphur 132
Cockroach 87
Cockroaches 30
Coenagrionidae 9-10
Coenobita compressus 184
Coenobitidae 184
Coffee family 26, 73, 105, 115-117, 136, 161, 173
Coleoptera 52-76
Coliadinae 129-135
Colobura dirce 168
Condylostylus 108

Conocephalinae 17
Convolvulaceae 69, 71
Copiphora cultricornis 17
Coreidae 40-41
Corydalidae 32
Corydalus 32
Costa Rican Redleg Tarantula 178
Crabronidae 88-89
Crabs 184-188
Crackers 163
Cramer's Eighty-eight 165
Creon Skipper 127
Creonpyge creon 127
Crimson Patch 172
Crustacea 184-188
Cucurbitaceae 137
Curculionidae 72-76
Cyanopepla scintillans 121
Cycads 144
Cyphomyia 106
Cyrestinae 167

Daggerwings 167
Damselflies 8-10
Danaus plexippus 145
Deep-blue Eyed-metalmark 136
Deer fly 88
Defensive compounds (see Chemical defenses)
Dermaptera 15
Diaethria clymena 165
Diaethria pandama 166
Diapheromeridae 24-26
Dichapetalaceae 120
Dinoe juno 157
Diogmites 107
Dione moneta 157
Diptera 102-112
Dirce Beauty 167
Disjunct Yellow 134
Dobsonflies 32
Dogbane family 121, 145-146
Dolichopodidae 108
Dorantes Longtail 125
Dragonflies 2, 11-14
Druce's Tiger Moth 120
Dryas iulia 158
Dryophthoridae 72
Dynastes septentrionalis 54
Dynastinae 54-55

Earwigs 15
Eciton burchellii 94-95
Edessa tauriformis 38
Edward's Bark Scorpion 177
Eggs (photos of) 3, 25-27, 72, 85
Elateridae ii, 58-59
Elephant Beetle 54
Emerald Cicada 44
Entiminae 73

Epigomphus tumefactus 11
Epirhyssa mexicana 78
Erato Longwing 159
Erebidae 119-121
Euchroma gigantea 57
Euglossa 90-91
Eumaeus godarti 144
Eumastacidae 20
Eumorpha anchemolus 1
Eumorpha vitis 117
Euphorbiaceae 39, 103, 118, 163
Eurema arbela 133-134
Eurema xantochlora 135
Eurhinus magnificus 73
Eurybia lycisca 137
Eusocial insects 30-31, 78, 85-86, 92-100
Exophthalmus jekelianus 73
Exophthalmus nicaraguensis 73
Eyes 2

Fabaceae 42, 48, 96, 104, 118-119, 125-
    126, 129, 132, 140-141, 152
Ferns 20
Ficus 82, 167
Fiddler Crab 185
Fig 82, 167
Fig Wasp 82
Fireflies 59-60
Flies 102-111
Flower flies 109
Formicidae 49, 93-100, 137, 141, 156
Four-spotted Eighty-eight 164
Fulgora lampetis 42
Fulgoridae 42-43
Fungi 15, 63-64, 98-99, 102, 152, 176

Gall midges 102-105
Galls 74, 102-105
Gasteracantha cancriformis 180
Gecarcinidae 188
Gecarcinus quadratus 188
Ghost crabs 186-187
Giant earwig 15
Giant Helicopter Damselfly 10
Giant Metallic Ceiba Borer 57
Glasswing butterflies 146-149
Godyris nero 146
Golden Carpenter Ant 100
Golden Silk Orbweaver 80, 181
Golden Target Beetle 70
Golden Tortoise Beetle 69
Gomphidae 11
Grape family 74
Grasses 121, 128
Grasshoppers 20-23
Great Kite-swallowtail 122
Green Banana Cockroach 30
Greta morgane 147
Gulf Fritillary 156

Gulf Ghost Crab 186-187
Hairstreaks 142-144
Halloween Crab 188
Hamadryas amphinome 163
Hanno Blue 141
Harlequin Longhorn Beetle 64
Harvestmen 176
Hawk moths 81, 116-117
Hecale Longwing 160
Helenor Blue Morpho 152-153
Heliconia (plant) 41, 45, 47, 76, 154-155
Heliconia Bug 41
Heliconia Spittlebug 45
Heliconiinae 156-160
Heliconius charithonia 158
Heliconius clysonymus 159
Heliconius erato 159
Heliconius hecale 160
Hemiargus hanno 141
Hemiptera 35-50
Heraclides androgeus 123
Hercules Beetle 54
Hermit crabs 184
Hesperiidae 125-128
Hetaerina occisa 8
Heteroptera 35-41
Homeomastax surda 20
Hoplocypode occidentalis 186-187
Hoplopyga liturata 190
Horned Mantis 29
Hornworm 116-117
Horse Guard Wasp 88
Hydropsychidae 33
Hymenoepimecis robertsae 80
Hymenoptera 78-100
Hypocrita drucei 120
Hystricia micans 111

Ichneumonidae 78-81
Inkblot Beetle 66
Ironclad Beetle 63
Ischnocodia annulus 70
Ithomiini 146-149

Jekel's Broad-nosed Weevil 73
Jewel Scarab 56
Jewel Wasp 87
Julia 158
Juno Silverspot 157

Katydids x, 5, 16-19
Kissing Bug 37

Ladoffa 46
Lampyridae 59-60
Lauraceae 150
Leaf beetles 66-71
Leaf butterflies 150-151
Leafcutter Ant 98-99
Leaf-footed bugs 40-41

Leafhoppers 46-48
Legume family 42, 96, 104, 118-119, 125-126, 129, 132, 135, 140-141, 152
Lepidoptera 113-174
Leptinotarsa undecimlineata 67
Leptonema 33
Leptoscelis tricolor 41
Leucauge mariana 182
Libellula herculea 12
Libellulidae 12-14
Lichens 27-28, 64
Limenitidinae 161-162
Lissomus ii
Liturgusa maya 27
Liturgusidae 27
Liverworts 28, 138
Locusts 22
Longhorn beetles 64-65
Long-legged Fly 108
Lowland Knobtail 11
Lycaenidae 1141-144
Lycidae 61

Mahanarva costaricensis 45
Mallophora 107
Mallow family 57, 66, 104
Malvaceae 57, 66, 104
Mantidae 28-29
Mantises 27-29
Mantispidae 192
Marantaceae 137, 154
Mariola Bee 92
Marpesia marcella 167
Mechanitis polymnia 149
Megaloprepus caerulatus 10
Megaloptera 32
Megaphobema mesomelas 178
Megasoma elephas 45
Melanis pixe 140
Meliponini 92
Melitaeini 172-174
Membracidae 49-50
Membracis mexicana 49
Mesoconius nigrihumeralis 110
Mesosemia asa 136
Messatoporus 79
Metallic Green Hoverfly 109
Metalmark butterflies 136-140
Mexican Silverspot 157
Mexican Treehopper 49
Micrathena 180
Micropezidae 110
Migration 94, 118, 129, 145, 147-148
Milkweed 145
Millipedes 183
Mimetica mortuifolia 16
Mimosa Yellow 132
Minuca mordax 185
Mischocyttarus basimacula 85
Mischocyttarus tolensis, 85

Molytinae 76
Monarch Butterfly 145
Monkey Grasshopper 20
Montane Longwing 159
Moraceae 82, 167
Morning glory family 69, 71
Morpho helenor 3, 152-153
Moss Stick Insect 24-25
Mosses 18, 24, 138
Moths 113-121
Mud dauber 74, 89
Muntingiaceae 127
Mutillidae 83
Myriopoda 183
Myrtaceae 105

Nasute Termite 31
Nasutitermes corniger 31
Neotheronia 79
Nephila 80, 181
Nero Clearwing 146
Nests 15, 31, 36, 78-79, 83, 85-101
Netelia 81
Net-spinning Caddisfly 33
Net-winged Beetle 61
Neuroptera 192
Nightshade family 21, 67, 142, 146-149
Number butterflies 164-166
Nymphalidae 3, 6, 145-174
Nymphalinae 168-174
Nymphalini 168-169
Nyssodesmus python 183

Ocypode gaudichaudi 186
Ocypodidae 185-187
Odonata 2, 8-14
Omaspides convexicollis 71
Oncometopia clarior 48
Orange-patched Crescent 173
Orange-striped Eighty-eight 166
Orchard Orbweaver 182
Orchid Bees 90-91
Orchids 24, 90
Organ Pipe Mud Dauber 89
Orizaba Silkmoth 1115
Ornidia obesa 109
Orthemis ferruginea 13
Orthoptera 16-23
Owl Moth 113-114

Pachycoris torridus 39
Pacific Hermit Crab 184
Painted Ghost Crab 186
Palms 68, 72, 154, 176
Panchlora nivea 30
Paper wasp 80, 85-86, 94
Papilionidae 122-124
Parandra 65
Paraponera clavata 93

Parasitic insects (parasitoids) 71, 78-81, 83-84, 87, 98, 111
Paromenia 47
Pasion flower/fruit/vine 40, 84, 156-157
Passalidae 53
Passiflora 40, 84, 156-157
Passion-vine butterflies 156-160
Peanut-headed Bug 42
Pegoscapus silvestrii 82
Pentatomidae 38
Pepsis aquila 84
Peridinetus cretaceus 75
Perithemis tenera 14
Phaneropterinae viii, 19
Phanocles costaricensis 26
Phanocles Stick Insect 26
Phasmatodea 24-26
Phoebis argante 130
Phoebis marcellina 132
Phoebis virgo 131
Photinus 60
Photuris crassa 60
Pierella helvina 6
Pieridae 129-135
Pimpla ichneumoniformis 80
Piper 75
Pipevine 124
Planthoppers 42-43
Platygonia praestantior 47
Poaceae 121, 128
Poanes zabulon 128
Poecilotylus species 110
Polistinae 80, 85-86, 94
Polybia emaciata 86
Polydamus Swallowtail 124
Polymnia Tigerwing 149
Polyommatinae 141
Pompilidae 84
Pond Amberwing 14
Prionostemma 176
Proctolabine Grasshopper 21
Protesilaus protesilaus 123
Pseudomethoca chontalensis 83
Pseudomyrmex flavicornis 96
Pseudomyrmex spinicola 96
Pseudophyllinae 18
Pseudoscorpions 64, 176
Pseudoxycheila tarsalis 52
Pterochrozinae 16
Pterodictya reticularis 43
Pteronymia simplex 148
Pupa (photos of) 3, 135, 148
Purple-stained Daggerwing 167
Pyrisitia nise 132
Pyrisitia proterpia 133
Pyrophorus 59
Python Millipede 183

Queen 31, 78, 85-86, 92-100

Red Cracker 163
Red-bordered Pixie 140
Red-eyed Argia 9
Red-washed Satyr 6
Reduviidae 35-37
Reticulated Lubber Grasshopper 23
Rhinostomus barbirostris 72
Rhionaeschna cornigera 2
Rhionaeschna psilus 11
Riodinidae 136-140
Robber Fly 107
Romaleidae 23
Roseate Skimmer 13
Rothschildia orizaba 115
Rubiaceae 26, 73, 105-117, 136, 161, 173
Rubyspot Damselfly 8
Rusty-tipped page 171
Rutaceae 43, 123
Rutelinae 56

Sapindaceae 164, 166
Sarota chrysus 138
Saturniidae 111, 113-115
Satyrinae 6, 152-155
Scarabaeidae viii, 54-56, 190
Sceliphron 79
Schistocerca centralis 22
Scorpions 177
Scutellaridae 39
Semiotus Click Beetle 58
Semiotus illigeri 58
Sharpshooter leafhoppers 46-48
Shield Bug 39
Shield Mantis 28
Silk 33, 146, 178-182
Silk moths 111, 113-115
Silver-sided Skimmer 12
Simaroubaceae 42
Simple Clearwing 148
Singing 16, 44
Siproeta epaphus 171
Sisters 161-162
Sixola Metalmark 139
Skippers 125-128
Smilacaceae 24-25
Smooth-banded Sister 161
Smyrna blomfildia 169
Snails 60, 184
Soapberry family 164, 166
Solanaceae 21, 67, 142, 146-149
Soldier Beetle 62
Soldier Fly 106
Sounds 16, 23, 44, 163
Soursop family 122
Sphingidae 1, 81, 116-117
Spiders 10, 62, 78, 84, 89, 178-182
Spinybacked Orbweaver 180
Spiracles 1-2, 3, 23

Spittle bugs 45
Splendid Weevil 74
Spurge family 39, 103, 118, 163
Stick insects 24-26
Stictia signata 88
Stilpnochlora azteca 19
Stilt-legged Fly 110
Sting 83-88, 93-94, 96, 113, 177
Stingless bees 36, 92
Stoll's Sarota 138
Strand's Groundstreak 143
Stratiomyiidae 106
Sulfur butterflies 129-135
Swallowtail butterflies 122-124
Sylvan Leaf Katydid 16
Syrphidae 109

Tachinidae 111
Taeniopoda reticulata 23
Tailed Orange 133
Tailed Sulphur 131
Tarantula Hawk 84
Tarantulas 84, 178
Tegosa anieta 174
Telegonus fulgerator complex 126
Tengella radiata 179
Tengella Spider 179
Termites 31
Termitidae 31
Tetragnathidae 182
Tetragonisca angustula 92
Tettigoniidae x, 5, 16-19
Theclinae 142-144
Theraphosidae 84, 178
Thick-tipped Greta 147
Thorn Bug 50
Thyreodon laticinctus 81
Tiger beetles 52
Tiger moths 111, 120-121
Togarna Hairstreak 142
Tortoise beetles 68-73
Tracta Sister 162
Treehoppers 49-60
Trema 165
Triatoma dimidiata 37
Trichonephila clavipes 80, 181
Trichoptera 33
Tropical Yellow 135
True bugs 35-41
Trychopeplus laciniatus 24-25
Trypanosoma cruzi 37

Trypoxylon monteverdeae 89
Turquoise-banded Shoemaker 150-151
Turquoise-tipped Darner 11
Two-barred Flasher 126
Two-spotted Tiger Beetle 52
Ultrasounds 115-116
Umbonia crassicornis 50
Urania fulgens 118
Urania Swallowtail Moth 118
Uraniidae 118
Urbanus viterboana 125
Urticaceae (see also *Cecropia*) 169

Vachellia collinsii 96
Vates pectinicornis 29
Vatinae 29
Velvet Ant 83
Verbenaceae 91, 123-125, 129-131, 156,
    158, 159
Vespidae 80, 85-86
Veturius sinuaticollis 53
Viburnum 162
Victorini 170-171
Vitaceae 74
Vochysiaceae 44

Warning coloration 39, 120, 156
Wasps 78-89
Wax-tailed Planthopper 43
Weevils 72-76
Wheel Bug 35
White Angled-sulphur 129
White-tipped Cycadian 144
Wireworm 58

Xylophanes crotonis 116
Xyophanes species 117
Yellow Angled-sulphur 130

Yellow-edged Owl Butterfly 154-155
Yellow-faced Spear Bearer 17

Zabulon Skipper 128
Zamia 144
Zammara smaragdina 44
Zebra Longwing 158
Zombie 87
Zopheridae 63
Zopherus jansoni 63
Zoropsidae 179
Zygogramma violaceomaculata 66